LONDON MATHEMATICAL SOCIETY LECTURE ~~

Managing Editor: Professor N.J. Hitchin, Mathematical Institute, University of Oxford, ~~ _
St Giles, Oxford OX1 3LB, United Kingdom

The titles below are available from booksellers, or, in case of difficulty, from Cambridge
University Press.

London Mathematical Society Lecture Note Series. 292

A Quantum Groups Primer

Shahn Majid
Queen Mary, University of London

CAMBRIDGE
UNIVERSITY PRESS

PUBLISHED BY THE PRESS SYNDICATE OF THE UNIVERSITY OF CAMBRIDGE
The Pitt Building, Trumpington Street, Cambridge, United Kingdom

CAMBRIDGE UNIVERSITY PRESS
The Edinburgh Building, Cambridge CB2 2RU, UK
40 West 20th Street, New York, NY 10011-4211, USA
477 Williamstown Road, Port Melbourne, VIC 3207, Australia
Ruiz de Alarcón 13, 28014 Madrid, Spain
Dock House, The Waterfront, Cape Town 8001, South Africa

http://www.cambridge.org

First published 2002

Typeface Computer Modern 10/13. *System* LaTeX 2_ε [Typeset by the author]

A catalogue record of this book is available from the British Library

Library of Congress Cataloguing in Publication data

ISBN 0 521 01041 1 paperback

Transferred to digital printing 2003

For my friends

Contents

Preface

Hopf algebras or 'quantum groups' are natural generalisations of groups. They have many remarkable properties and, nowadays, they come with a wealth of examples and applications in pure mathematics and mathematical physics.

Most important are the quantum groups $U_q(\mathfrak{g})$ modelled on, and in some ways more natural than, the enveloping algebras $U(\mathfrak{g})$ of simple Lie algebras $\mathfrak{g}$. They provide a natural extension of Lie theory. There are also finite-dimensional quantum groups such as bicrossproduct quantum groups associated to the factorisation of finite groups. Moreover, quantum groups are clearly indicative of a more general 'noncommutative geometry' in which coordinate rings are allowed to be noncommutative algebras.

This is a self-contained first introduction to quantum groups as algebraic objects. It should also be useful to someone primarily interested in algebraic groups, knot theory or (on the mathematical physics side) q-deformed physics, integrable systems, or conformal field theory. The only prerequisites are basic algebra and linear algebra. Some exposure to semisimple Lie algebras will also be useful.

The approach is basically that taken in my 1995 textbook, to which the present work can be viewed as a companion 'primer' for pure mathematicians. As such it should be a useful complement to that much longer text (which was written for a wide audience including theoretical physicists). In addition, I have included more advanced topics taken from my review on Hopf algebras in braided categories and subsequent research papers given in the Bibliography, notably the 'braided geometry' of $U_q(\mathfrak{g})$. This is material which may eventually be developed in a sequel volume to the 1995 text.

In particular, our approach differs significantly from that in other textbooks on quantum groups in that we do not define $U_q(\mathfrak{g})$ by means of generators and relations 'pulled out of a hat' but rather we deduce these from a more conceptual braided-categorical construction. Among the benefits of this approach is an inductive definition of $U_q(\mathfrak{g})$ as given by the repeated adjunction of 'quantum planes'. The latter, as well as the subalgebras $U_q(n_+)$, are constructed in our approach as braided groups, which can be viewed as a modern braided-categorical setting for the first (easy) part of Lusztig's text.

The book itself is the verbatim text of a course of 24 lectures on *Quantum Groups* given in the Department of Pure Mathematics and Mathematical Statistics at the University of Cambridge in the Spring of 1998. The course was at the *Part III* diploma level of the mathematics tripos, which is approximately the level of a first year graduate course at an American university, perhaps a bit less advanced. Accordingly, it should be possible to base a similar course on this book, for which purpose I have retained the original lecture numbering. The first 1/3 of the lectures cover the basic algebraic structure, the second 1/3 the representation theory and the last 1/3 more advanced topics. There were also three useful problem sets distributed during the course, which I include at the end of the book.

I would like to thank the students who attended the course for their useful comments. Particularly, the lectures start off quite slowly with a lot of explicit computations and notations from the theory of Hopf algebras; depending on the wishes of the students, one could skip faster through these lectures by deferring the proofs as exercises – with solutions on handouts. Meanwhile, the last five lectures are an introduction to some miscellaneous topics; they are self-contained and could be omitted, depending on the time available. Finally, I want to thank Pembroke College in the University of Cambridge, where I was based at the time and during much of the period of writing.

Shahn Majid
School of Mathematical Sciences
Queen Mary, University of London

1

Coalgebras, bialgebras and Hopf algebras. $U_q(b_+)$

Quantum groups today are like groups were in the nineteenth century, by which I mean

– a young theory, abundant examples, a rich and beautiful mathematical structure. By 'young' I mean that many problems remain wide open, for example the classification of finite-dimensional quantum groups.

– a clear need for something *like* this in the mathematical physics of the day. In our case it means quantum theory, which clearly suggests the need for some kind of 'quantum geometry', of which quantum groups would be the group objects.

These are *algebra* lectures, so we will not be able to say too much about physics. Suffice it to say that the familiar 'geometrical' picture for classical mechanics: symplectic structures, Riemannian geometry, is all thrown away when we look at quantum systems. In quantum systems the classical variables or 'coordinates' are replaced by operators on a Hilbert space and typically generate a noncommutative algebra, instead of a commutative coordinate ring as in the classical case. There is a need for geometrical structures on such quantum systems parallel to those in the classical case. This is needed if geometrical ideas such as gravity are ever to be unified with quantum theory.

From a mathematical point of view, the motivation for quantum groups is:

– the original (dim) origins in cohomology of groups (H. Hopf, 1947); an older name for quantum groups is 'Hopf algebras'

– q-deformed enveloping algebra quantum groups provide an explanation for the theory of q-special functions, which dates back to the 1900s. They are used also in number theory. (For example, there are q-exponentials etc., related to quantum groups as ordinary exponentials are related to the additive group $\mathbb{R}$.)

1

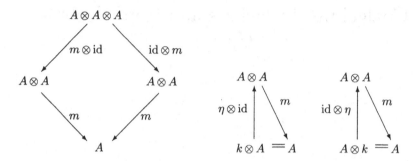

Fig. 1.1. Associativity and unit element expressed as commutative diagrams.

 – representations of quantum groups form braided categories, leading to link invariants
 – quantum groups are the 'group' objects in some kind of noncommutative algebraic geometry
 – quantum groups are the 'transformation' objects in noncommutative algebraic geometry
 – quantum groups restore an input–output symmetry to algebraic constructions; for example, they admit Fourier theory.

We fix a field k over which we work. We begin by recalling that an algebra A is
1. A vector space over k.
2. A map $m : A \otimes A \to A$ which is associative in the sense $(ab)c = a(bc)$ for all $a, b, c \in A$. Here $ab = m(a \otimes b)$ is shorthand.
3. A unit element 1_A, which we write equivalently as a map $\eta : k \to A$ by $\eta(1) = 1_A$. We require $a 1_A = a = 1_A a$ for all $a \in A$.
In terms of the maps, these axioms are given by the commutative diagrams in Figure 1.1. Note that most algebraic constructions can, like the axioms themselves, be expressed as commuting diagrams. When all premises, statements and proofs of a theorem are written out like this then reversing all arrows will also yield the premises, statements and proofs of a different theorem, called the 'dual theorem'.

Definition 1.1 *A coalgebra C is*
 1. A vector space over k.
 2. A map $\Delta : C \to C \otimes C$ (the 'coproduct') which is coassociative in

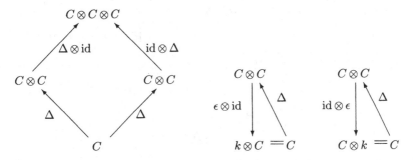

Fig. 1.2. Coassociativity and counit element expressed as commutative diagrams.

the sense

$$\sum c_{(1)(1)} \otimes c_{(1)(2)} \otimes c_{(2)} = \sum c_{(1)} \otimes c_{(2)(1)} \otimes c_{(2)(2)}$$

for all $c \in C$. Here $\Delta c \equiv \sum c_{(1)} \otimes c_{(2)}$ is shorthand.

3. A map $\epsilon : C \to k$ (the 'counit') obeying $\sum \epsilon(c_{(1)})c_{(2)} = c = \sum c_{(1)}\epsilon(c_{(2)})$ for all $c \in C$.

In terms of the maps, these axioms are given by the commutative diagrams in Figure 1.2, which is just Figure 1.1 with all arrows reversed.

This notion of reversing arrows has the same status as the idea, familiar in algebra, of having both left and right module versions of a construction. The theory with only left modules is equivalent to the theory with right modules, by a left–right reflection (i.e. reversal of tensor product). But one can also consider theorems with both left and right modules interacting in some way, e.g. bimodules. Similarly, the arrow-reversal operation transforms theorems about algebras to theorems about coalgebras. However, we can also consider theorems involving both concepts. In this way, quantum group theory is a very natural 'completion' of algebra to a setting which is invariant under the arrow-reversal operation.

Definition 1.2 *A bialgebra H is*

1. An algebra H, m, η.

2. A coalgebra H, Δ, ϵ.

3. Δ, ϵ are algebra maps, where $H \otimes H$ has the tensor product algebra structure $(h \otimes g)(h' \otimes g') = hh' \otimes gg'$ for all $h, h', g, g' \in H$.

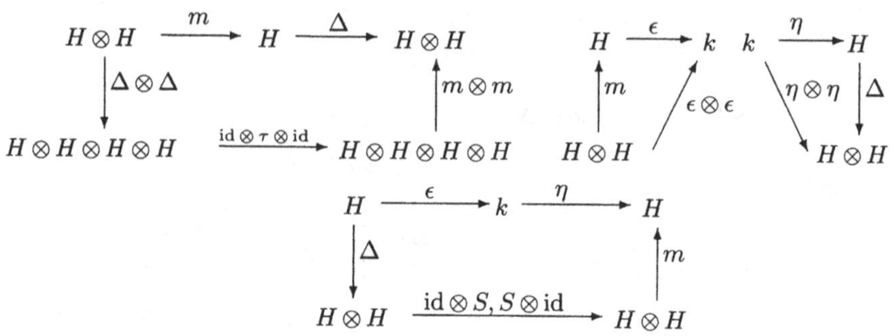

Fig. 1.3. Additional axioms that make the algebra and coalgebra H into a Hopf algebra.

Actually, a bialgebra is more like a quantum 'semigroup'. We need something playing the role of group inversion:

Definition 1.3 *A Hopf algebra H is*

1. A bialgebra $H, \Delta, \epsilon, m, \eta$.

2. A map $S : H \to H$ (the 'antipode') such that $\sum (Sh_{(1)})h_{(2)} = \epsilon(h) = \sum h_{(1)} Sh_{(2)}$ for all $h \in H$.

The axioms that make a simultaneous algebra and coalgebra into a Hopf algebra are shown in Figure 1.3, where $\tau : H \otimes H \to H \otimes H$ is the 'flip' map $\tau(h \otimes g) = g \otimes h$ for all $h, g \in H$.

Proposition 1.4 *(Antihomomorphism property of antipodes). The antipode of a Hopf algebra is unique and obeys $S(hg) = S(g)S(h)$, $S(1) = 1$ (i.e. S is an* antialgebra *map) and $(S \otimes S) \circ \Delta h = \tau \circ \Delta \circ Sh$, $\epsilon Sh = \epsilon h$ (i.e. S is an* anticoalgebra *map), for all $h, g \in H$.*

Proof During proofs, we will usually omit the $\sum$ signs, which should be understood. If S, S_1 are two antipodes on a bialgebra H then they are equal because $S_1 h = (S_1 h_{(1)}) \epsilon(h_{(2)}) = (S_1 h_{(1)}) h_{(2)(1)} Sh_{(2)(2)} = (S_1 h_{(1)(1)}) h_{(1)(2)} Sh_{(2)} = \epsilon(h_{(1)}) Sh_{(2)} = Sh$. Here we wrote $h = h_{(1)} \epsilon(h_{(2)})$ by the counit axioms, and then inserted $h_{(2)(1)} Sh_{(2)(2)}$ knowing that it would collapse to $\epsilon(h_{(2)})$. We then used associativity and (the more novel ingredient) coassociativity to be able to collapse $(S_1 h_{(1)(1)}) h_{(1)(2)}$ to $\epsilon(h_{(1)})$. Note that the proof is not any harder than the usual one for uniqueness of group inverses, the only complication being that we are working now with parts of linear combinations and have to take care to keep the order of the coproducts. We can similarly collapse such expressions as $(S_1 h_{(1)})h_{(2)}$ or $h_{(2)}Sh_{(3)}$ wherever they occur as long as the two collapsing factors are in linear order. This is just the analogue of cancelling $h^{-1}h$ or hh^{-1} in a group. Armed with such techniques, we return now to the proof of the proposition. Consider the identity

$$(S(h_{(1)(1)}g_{(1)(1)}))h_{(1)(2)}g_{(1)(2)} \otimes g_{(2)} \otimes h_{(2)}$$
$$= (S((h_{(1)}g_{(1)})_{(1)}))(h_{(1)}g_{(1)})_{(2)} \otimes g_{(2)} \otimes h_{(2)}$$
$$= \epsilon(h_{(1)}g_{(1)})1 \otimes g_{(2)} \otimes h_{(2)} = 1 \otimes g \otimes h.$$

We used that Δ is an algebra homomorphism, then the antipode axiom applied to $h_{(1)}g_{(1)}$. Then we used the counity axiom. Now apply S to the middle factor of both sides and multiply the first two factors. One has the identity

$$Sg \otimes h = (S(h_{(1)(1)}g_{(1)(1)}))h_{(1)(2)}g_{(1)(2)}Sg_{(2)} \otimes h_{(2)}$$
$$= (S(h_{(1)(1)}g_{(1)}))h_{(1)(2)}g_{(2)(1)}Sg_{(2)(2)} \otimes h_{(2)} = (S(h_{(1)(1)}g))h_{(1)(2)} \otimes h_{(2)},$$

where we used coassociativity applied to g. We then use the antipode axiom applied to $g_{(2)}$, and the counity axiom. We now apply S to the second factor and multiply up, to give

$$(Sg)(Sh) = (S(h_{(1)(1)}g))h_{(1)(2)}Sh_{(2)} = (S(h_{(1)}g))h_{(2)(1)}Sh_{(2)(2)} = S(hg).$$

We used coassociativity applied to h, followed by the antipode axioms applied to $h_{(2)}$ and the counity axiom. $\qquad\square$

Example 1.5 *The Hopf algebra $H = U_q(b_+)$ is generated by 1 and the elements X, g, g^{-1} with relations*

$$gg^{-1} = 1 = g^{-1}g, \quad gX = qXg,$$

where q is a fixed invertible element of the field k. Here

$$\Delta X = X \otimes 1 + g \otimes X, \quad \Delta g = g \otimes g, \quad \Delta g^{-1} = g^{-1} \otimes g^{-1},$$

$$\epsilon X = 0, \quad \epsilon g = 1 = \epsilon g^{-1}, \quad SX = -g^{-1}X, \quad Sg = g^{-1}, \quad Sg^{-1} = g.$$

Note that $S^2 \neq \mathrm{id}$ in this example (because $S^2 X = q^{-1}X$).

Proof We have Δ, ϵ on the generators and extend them multiplicatively to products of the generators (so that they are necessarily algebra maps as required). However, we have to check that this is consistent with the relations in the algebra. For example, $\Delta gX = (\Delta g)(\Delta X) = (g \otimes g)(X \otimes 1 + g \otimes X) = gX \otimes g + g^2 \otimes gX$, while equal to this must be $\Delta qXg = q(\Delta X)(\Delta g) = q(X \otimes 1 + g \otimes X)(g \otimes g) = qXg \otimes g + qg^2 \otimes Xg$. These expressions are equal, using again the relations in the algebra as stated. Similarly for the other relations. For the antipode, we keep in mind the preceding proposition and extend S as an antialgebra map, and check that this is consistent in the same way. Since S obeys the antipode axioms on the generators (an easy computation), it follows that it obeys them also on the products since Δ, ϵ are already extended multiplicatively. $\square$

It is a nice exercise – we will prove it later in the course, but some readers may want to have fun doing it now – to show that

$$\Delta X^m = \sum_{r=0}^{m} \begin{bmatrix} m \\ r \end{bmatrix}_q X^{m-r} g^r \otimes X^r$$

where

$$\begin{bmatrix} m \\ r \end{bmatrix}_q = \frac{[m]_q!}{[r]_q![m-r]_q!}, \quad [r]_q! = [r]_q[r-1]_q \cdots [1]_q$$

are the q-binomial coefficients defined in terms of 'q-integers'

$$[r]_q = 1 + q + \cdots + q^{r-1} = \frac{1 - q^r}{1 - q}.$$

The last expression here should be used only when $q \neq 1$, of course. We should also assume $[r]_q$ are invertible to write the q-binomial coefficients in this way.

Example 1.6 *Let G be a finite group. The* group Hopf algebra kG *is the vector space with basis G, and the algebra structure, unit, coproduct, counit and antipode*

$$\text{product in } G, \quad 1 = e, \quad \Delta g = g \otimes g, \quad \epsilon g = 1, \quad Sg = g^{-1}$$

on the basis elements $g \in G$ (extended by linearity to all of kG).

Proof The multiplication is clearly associative because the group multiplication is. The coproduct is coassociative because it is so on each of the basis elements $g \in G$. It is an algebra homomorphism because $\Delta(gh) = gh \otimes gh = (g \otimes g)(h \otimes h) = (\Delta g)(\Delta h)$. The other facts are equally easy. G does not actually need to be finite for this construction, but we will be interested in the finite case. $\square$

So all of finite group theory should, in principle, be a special case of Hopf algebra theory. The same is true for Lie theory, if we use the enveloping algebra. We recall that a Lie algebra is:

1. A vector space $\mathfrak{g}$.
2. A map $[\ ,\ ] : \mathfrak{g} \otimes \mathfrak{g} \to \mathfrak{g}$ obeying the Jacobi identity and antisymmetry axioms (when the characteristic of k is not 2).

Example 1.7 *Let $\mathfrak{g}$ be a finite-dimensional Lie algebra over k. The universal enveloping Hopf algebra $U(\mathfrak{g})$ is the noncommutative algebra generated by 1 and elements of a basis of $\mathfrak{g}$ modulo the relations $[\xi, \eta] = \xi\eta - \eta\xi$ for all ξ, η in the basis. The coproduct, counit and antipode are*

$$\Delta\xi = \xi \otimes 1 + 1 \otimes \xi, \quad \epsilon\xi = 0, \quad S\xi = -\xi$$

extended in the case of Δ, ϵ as algebra maps, and in the case of S as an antialgebra map.

Proof We extend Δ, ϵ as algebra homomorphisms and S as an antialgebra homomorphism, and have to check that this extension is consistent with the relations. For example, $\Delta(\xi\eta) = (\xi \otimes 1 + 1 \otimes \xi)(\eta \otimes 1 + 1 \otimes \eta) = \xi\eta \otimes 1 + 1 \otimes \xi\eta + \xi \otimes \eta + \eta \otimes \xi$. Subtracting from this the corresponding expression for $\Delta\eta\xi$ and using the relations, we obtain $[\xi, \eta] \otimes 1 + 1 \otimes [\xi, \eta] = \Delta[\xi, \eta]$ as required. Similarly for the counit and antipode. $\square$

One can say, informally, that $U(\mathfrak{g})$ is generated by 1 and elements of $\mathfrak{g}$ with the relations stated; it does not depend on a choice of basis. A more formal way to say this is to construct first the tensor Hopf algebra $T(V) = k \oplus V \oplus V \otimes V \oplus V \otimes V \otimes V \oplus \cdots$ on any vector space V. The product here is $(v \otimes \cdots \otimes w)(x \otimes \cdots \otimes y) = (v \otimes \cdots \otimes w \otimes x \otimes \cdots \otimes y)$. This forms a Hopf algebra with

$$\Delta v = v \otimes 1 + 1 \otimes v, \quad \epsilon v = 0, \quad Sv = -v$$

for all $v \in V$. The enveloping algebra $U(\mathfrak{g})$ is the quotient of $T(\mathfrak{g})$ modulo the ideal generated by the relations $\xi \otimes \eta - \eta \otimes \xi = [\xi, \eta]$. (Of course, the best definition is as a universal object, but we will not need that.)

So Lie theory is also contained, in principle, as a special case of quantum group theory. In fact, one of the main motivations for Hopf algebras in the 1960s was precisely as a tool that unifies the treatment of results for groups and Lie algebras into one technology, e.g. their cohomology theory. Clearly, our example $U_q(b_+)$ is a mixture of these two kinds of 'classical' Hopf algebras. It has an element g which is *grouplike* in the sense that it obeys $\Delta g = g \otimes g$. And it has an element X which is a bit like the Lie case. But it is neither a group algebra nor an enveloping algebra exactly. What characterises these classical objects, in contrast to $U_q(b_+)$, is:

Definition 1.8 *A Hopf algebra is commutative if it is commutative as an algebra. It is 'cocommutative' if it is cocommutative as a coalgebra, i.e. if $\tau \circ \Delta = \Delta$. This is the arrows-reversed version of commutativity.*

Corollary 1.9 *If H is a commutative or cocommutative Hopf algebra, then $S^2 = $ id.*

Proof We use Proposition 1.4, so that $S^2h = (S^2h_{(1)})(Sh_{(2)})h_{(3)} = (S(h_{(1)}Sh_{(2)}))h_{(3)} = h$ in the cocommutative case. Here we use a neutral notation $h_{(1)} \otimes h_{(2)} \otimes h_{(3)} \equiv h_{(1)(1)} \otimes h_{(1)(2)} \otimes h_{(2)} = h_{(1)} \otimes h_{(2)(1)} \otimes h_{(2)(2)}$ (just as one writes $abc \equiv (ab)c = a(bc)$). The other case is similar.   □

Clearly, kG and $U(\mathfrak{g})$ are cocommutative. The coordinate rings of linear algebraic groups are likewise commutative Hopf algebras, while $U_q(b_+)$ is neither. As a tentative definition, we can say that a truly 'quantum' group (in contrast to a classical group or Lie object viewed as one) is a noncommutative and noncocommutative Hopf algebra. Later on, we will add further properties as well.

2
Dual pairing. $SL_q(2)$. Actions

In the last lecture we showed how to view finite groups and Lie algebras as Hopf algebras, and gave a variant that was truly 'quantum'. We now complete our basic collection of examples with some other classical objects.

Example 2.1 *Let G be a finite group with identity e. The* group function Hopf algebra $k(G)$ *is the algebra of functions on G with values in k and the pointwise product $(fg)(x) = f(x)g(x)$ for all $x \in G$ and $f, g \in k(G)$. The coproduct, counit and antipode are*

$$(\Delta f)(x, y) = f(xy), \quad \epsilon f = f(e), \quad (Sf)(x) = f(x^{-1}),$$

where we identify $k(G) \otimes k(G) = k(G \times G)$ (functions of two group variables).

Proof Coassociativity is evidently $((\Delta \otimes \mathrm{id})\Delta f)(x, y, z) = (\Delta f)(xy, z) = f((xy)z) = f(x(yz)) = (\Delta f)(x, yz) = ((\mathrm{id} \otimes \Delta)\Delta f)(x, y, z)$. Note that it comes directly from associativity in the group. Likewise, the counity and antipode axioms come directly from the group axioms for the unit element and inverse. $\qquad\square$

Also, when g is a finite-dimensional complex semisimple Lie algebra (as classified by Dynkin diagrams), it has an associated complex Lie group $G \subset M_n(\mathbb{C})$ (the $n \times n$ matrices with values in $\mathbb{C}$). This subset is of the form $G = \{x \in M_n \mid p(x) = 0\}$, where p is a collection of polynomial equations. Correspondingly, we have an algebraic variety with coordinate algebra $\mathbb{C}[G]$ defined as $\mathbb{C}[x^i{}_j]$ where $i, j = 1, \ldots, n$ (polynomials in n^2 variables), modulo the ideal generated by the relations $p(x) = 0$. The group structure inherited from matrix multiplication

9

corresponds to a coproduct and counit

$$\Delta x^i{}_j = \sum_k x^i{}_k \otimes x^k{}_j, \quad \epsilon(x^i{}_j) = \delta^i{}_j$$

where $\delta^i{}_j$ is the Kronecker delta-function. There is also an antipode given algebraically via a matrix of cofactors of the matrix $x^i{}_j$ of generators. In this way, we have a complex linear algebraic group with coordinate algebra $\mathbb{C}[G]$ as a Hopf algebra. In fact, G can be taken so that the coefficients of $p(x)$ are integers (from work of Chevalley) giving a coordinate ring $\mathbb{Z}[G]$. Then, by tensoring with k, the same construction works over any field and provides a Hopf algebra $k[G]$ (and considering all k, one has an affine group scheme).

Example 2.2 *The Hopf algebra* $k[SL_2]$ *is* $k[a, b, c, d]$ *modulo the relation*

$$\det \begin{pmatrix} a & b \\ c & d \end{pmatrix} = 1.$$

The coproduct, counit and antipode are

$$\Delta a = a \otimes a + b \otimes c, \quad \Delta b = b \otimes d + a \otimes b, \quad \Delta c = c \otimes a + d \otimes c,$$

$$\Delta d = d \otimes d + c \otimes b, \quad \epsilon(a) = \epsilon(d) = 1, \quad \epsilon(b) = \epsilon(c) = 0,$$

$$Sa = d, \quad Sd = a, \quad Sb = c, \quad Sc = b.$$

The coalgebra and antipode here can be written more concisely as

$$\Delta \begin{pmatrix} a & b \\ c & d \end{pmatrix} = \begin{pmatrix} a & b \\ c & d \end{pmatrix} \otimes \begin{pmatrix} a & b \\ c & d \end{pmatrix}, \quad \epsilon \begin{pmatrix} a & b \\ c & d \end{pmatrix} = \begin{pmatrix} 1 & 0 \\ 0 & 1 \end{pmatrix},$$

$$S \begin{pmatrix} a & b \\ c & d \end{pmatrix} = \begin{pmatrix} d & -b \\ -c & a \end{pmatrix},$$

where matrix multiplication should be understood in this definition of Δ. This is no more than a shorthand notation. Finally, for a truly 'quantum' variant of this:

Example 2.3 *Let* $q \in k^*$. *The Hopf algebra* $SL_q(2)$ *is* $k\langle a, b, c, d \rangle$ *(the free associative algebra) modulo the ideal generated by the six 'q-*

commutativity' relations

$$ca = qac, \quad ba = qab, \quad db = qbd, \quad dc = qcd, \quad bc = cb,$$

$$da - ad = (q - q^{-1})bc$$

and the 'q-determinant' relation

$$ad - q^{-1}bc = 1.$$

The coalgebra has the same matrix form on the generators as above, and the antipode is

$$Sd = a, \quad Sa = d, \quad Sb = -qb, \quad Sc = -q^{-1}c.$$

Proof We will give general constructions for this kind of quantum group later on. For the moment, it is easy enough to verify directly that it fulfils the axioms. Hint: first consider the algebra $M_q(2)$ defined in the same way but without the q-determinant relation. This is a quadratic algebra and it is easier to verify that Δ, ϵ are well-defined when extended to products. Then show that $ad - q^{-1}bc$ is central in this algebra and grouplike in the sense $\Delta(ad - q^{-1}bc) = (ad - q^{-1}bc) \otimes (ad - q^{-1}bc)$ and $\epsilon(ad - q^{-1}bc) = 1$. The further relation $ad - q^{-1}bc = 1$ can then be added and the quotient remains a bialgebra. For the antipode, it is easy enough to see that it extends antimultiplicatively. Once well-defined, it is enough to check the antipode axioms on the generators, which is elementary. Note that in the compact 'matrix' notation one writes

$$ad - q^{-1}bc \equiv \det_q \begin{pmatrix} a & b \\ c & d \end{pmatrix}, \quad S\begin{pmatrix} a & b \\ c & d \end{pmatrix} = \begin{pmatrix} d & -qb \\ -q^{-1}c & a \end{pmatrix}.$$

$\square$

The quantum group $SL_q(2)$ here is also variously denoted $k_q[SL_2]$ or $\mathcal{O}_q(SL_2)$ in the literature. It completes our collection of basic examples. Here $kG, k(G)$ for finite groups and $U(\mathfrak{g}), k[G]$ for Lie algebras are 'classical' objects, while $U_q(b_+)$ and $SL_q(2)$ are more novel and truly 'quantum' groups according to the tentative definition given at the end of the last lecture. We now return to the general theory of Hopf algebras.

Definition 2.4 *Two Hopf algebras H, H' are 'dually paired' by a map $\langle\,,\,\rangle : H' \otimes H \to k$ if*

$$\langle \phi\psi, h \rangle = \langle \phi \otimes \psi, \Delta h \rangle, \quad \langle 1, h \rangle = \epsilon(h)$$

$$\langle \Delta\phi, h \otimes g \rangle = \langle \phi, hg \rangle, \quad \epsilon(\phi) = \langle \phi, 1 \rangle$$

$$\langle S\phi, h \rangle = \langle \phi, Sh \rangle$$

for all $\phi, \psi \in H'$ and $h, g \in H$. Here $\langle\,,\,\rangle$ extends to tensor products pairwise.

This says that the product of H and coproduct of H' are adjoint to each other under $\langle\,,\,\rangle$, and vice-versa. Likewise, the units and counits are mutually adjoint, and the antipodes are adjoint. The definition is made possible by the invariance of the Hopf algebra axioms under arrow-reversal (i.e. input–output symmetry) as explained in the last lecture.

If H is finite dimensional, then $\langle\,,\,\rangle = $ ev (the evaluation map) provides a duality pairing with H^*. Here, H^* has the product Δ^* and the coproduct m^*, where

$$\Delta^* : (H \otimes H)^* \to H^*, \quad m^* : H \to (H \otimes H)^*$$

are the duals of Δ, m of H. They define the required maps since $(H \otimes H)^* \supseteq H^* \otimes H^*$ is an equality for a finite-dimensional vector space H (otherwise, it need not be an equality and m^* need not descend to a coproduct on H^*). This is the unique possibility for a nondegenerate duality pairing in the finite-dimensional case, and we say H^* is the dual Hopf algebra in this case.

Among our examples, $k(G)^* = kG$ (by evaluation) and $U(\mathfrak{g}), \mathbb{C}[G]$ are dually paired over $\mathbb{C}$ for $\mathfrak{g}$ a finite-dimensional complex semisimple Lie algebra. If $\rho : \mathfrak{g} \subset M_n(\mathbb{C})$ is the defining representation of $\mathfrak{g}$, the pairing is

$$\langle \xi, x^i{}_j \rangle = \rho(\xi)^i{}_j, \quad \forall \xi \in \mathfrak{g}.$$

This result also extends to a general field with both $\mathfrak{g}$ and ρ defined over k. Meanwhile, $U_q(b_+)$ is self-dual:

Proposition 2.5 *$U_q(b_+)$ is dually paired with itself by*

$$\langle g, g \rangle = q, \quad \langle X, X \rangle = 1, \quad \langle X, g \rangle = \langle g, X \rangle = 0.$$

Proof We will see general methods for this kind of result later in the course. For the moment, it is a nice exercise directly from the definitions. Hint: first find that $f_{m,n}(g) \equiv \langle X^m g^n, g \rangle = \langle X^m, g \rangle \langle g^n, g \rangle = q^n \delta_{m,0}$ and $f_{m,n}(X) \equiv \langle X^m g^n, X \rangle = \langle X^m, X \rangle \langle g^n, 1 \rangle = \delta_{m,1}$. Then the coproduct $\Delta(X^m g^n) = (\Delta X^m)(g^n \otimes g^n)$ given in the last lecture, and the axioms of a pairing, imply that

$$f_{m,n}(hh') = \sum_{r=0}^{m} \begin{bmatrix} m \\ r \end{bmatrix}_q f_{m-r,n+r}(h) f_{r,n}(h')$$

for all $h, h' \in U_q(b_+)$. This determines $f_{m,n}$ on products, which shows that $\langle \ , \ \rangle$ is uniquely determined. We then define it on the basis $\{ X^m g^n | \ n \in \mathbb{Z}, \ m \in \mathbb{Z}_+ \}$ of each copy of $U_q(b_+)$ (where $\mathbb{Z}_+$ includes 0), by the resulting formula for f_{mn}, and verify the duality pairing axioms on products and coproducts of basis elements. $\qquad \square$

Finally, by definition, an action or representation of a bialgebra or Hopf algebra H means one of the underlying algebra. What is special about having a bialgebra is that one may tensor product representations. Clearly, if V, W are H-modules (i.e. H acts on them), then

$$h \triangleright (v \otimes w) = \sum h_{(1)} \triangleright v \otimes h_{(2)} \triangleright w \equiv (\Delta h) \triangleright (v \otimes w)$$

for all $h \in H$ and $v \in V, w \in W$, makes $V \otimes W$ into a H-module. Here $\triangleright$ is used to denote a left action. One always has a trivial module $V = k$, with

$$h \triangleright \lambda = \epsilon(h) \lambda, \quad \forall h \in H, \ \lambda \in k.$$

This is the identity object under the tensor product of modules.

Definition 2.6 *A bialgebra or Hopf algebra H acts on an algebra A (one says that A is an H-module algebra) if*

    1. H acts on A as a vector space.
    2. The product map $m : A \otimes A \to A$ commutes with the action of H.
    3. The unit map $\eta : k \to A$ commutes with the action of H.

Explicitly, the conditions 2,3 are

$$h \triangleright (ab) = \sum (h_{(1)} \triangleright a)(h_{(2)} \triangleright b), \quad h \triangleright 1 = \epsilon(h) 1, \quad \forall a, b \in A, \quad h \in H.$$

We leave it as an easy exercise to see what these conditions mean for our basic examples. One finds, for all $a, b \in A$:

    (i) for kG,

$$g \triangleright (ab) = (g \triangleright a)(g \triangleright b), \quad g \triangleright 1 = 1, \quad \forall g \in G,$$

which is the usual notion of a group action by automorphisms.

(ii) for $U(\mathfrak{g})$,

$$\xi\triangleright(ab) = (\xi\triangleright a)b + a(\xi\triangleright b), \quad \xi\triangleright 1 = 0, \quad \forall\xi \in \mathfrak{g}$$

which is the usual notion of a Lie action by derivations.

(iii) for $U_q(b_+)$,

$$g\triangleright(ab) = (g\triangleright a)(g\triangleright b), \quad X\triangleright(ab) = (X\triangleright a)b + (g\triangleright a)(X\triangleright b),$$

$$g\triangleright 1 = 1, \quad X\triangleright 1 = 0.$$

One says that X acts as a 'skew-derivation'.

(iv) for $k(G)$, it means A is a G-graded algebra, where $f\triangleright(a) = f(|a|)a$ on homogeneous elements of degree $|a|$. Here an action of $k(G)$ on a vector space V is the same thing as a G-grading $V = \bigoplus_{g\in G} V_g$, where we say that $|v| = g$ for all $v \in V_g$.

The situation for $k[SL_2]$ is roughly similar to (iv), but is not usually considered in any context that I know of; likewise for $SL_q(2)$.

Proposition 2.7 *(Adjoint action). Every Hopf algebra H acts on itself as an algebra by*

$$\mathrm{Ad}_h(g) = \sum h_{(1)}gSh_{(2)}$$

for all $h, g \in H$.

Proof We check $h\triangleright(g\triangleright a) = h\triangleright(g_{(1)}aSg_{(2)}) = h_{(1)}g_{(1)}a(Sg_{(2)})(Sh_{(2)}) = (hg)_{(1)}aS(hg)_{(2)} = (hg)\triangleright a$ using Proposition 1.4 about the antipode. Also, $1\triangleright a = 1aS(1) = a$. Thus, we have an action. We have a module algebra because $h\triangleright(ab) = h_{(1)}ab(Sh_{(2)}) = h_{(1)}a(Sh_{(2)})h_{(3)}bSh_{(4)} = (h_{(1)}\triangleright a)(h_{(2)}\triangleright b)$ and $h\triangleright 1 = h_{(1)}1Sh_{(2)} = \epsilon(h)$. We insert $(Sh_{(2)})h_{(3)}$, knowing that it collapses using the antipode axioms, and freely renumber to express coassociativity. Here $h_{(1)} \otimes h_{(2)} \otimes h_{(3)} \otimes h_{(4)}$ is our neutral notation denoting any of the five expressions $(\Delta \otimes \mathrm{id} \otimes \mathrm{id})(\Delta \otimes \mathrm{id})\Delta h$, $(\mathrm{id} \otimes \Delta \otimes \mathrm{id})(\Delta \otimes \mathrm{id})\Delta h$, etc. coinciding through coassociativity. $\qquad\square$

For our standard examples, we have (immediately from the definitions):

(i) for kG,

$$\mathrm{Ad}_g(h) = ghg^{-1}, \quad \forall g, h \in G.$$

(ii) for $U(\mathfrak{g})$,

$$\mathrm{Ad}_\xi(h) = \xi h - h\xi, \quad \forall \xi \in \mathfrak{g}, \ h \in U(\mathfrak{g}).$$

(iii) for $U_q(b_+)$,

$$\mathrm{Ad}_g(h) = q^{|h|}h, \quad \mathrm{Ad}_X(h) = Xh - q^{|h|}hX$$

for all $h \in U_q(b_+)$ of homogeneous degree $|h|$ in X.

(iv) for $k(G)$ and $k[G]$, the adjoint action is trivial because these algebras are commutative, so that Ad collapses by the antipode axioms.

(v) for $SL_q(2)$, one finds for example

$$\mathrm{Ad}_b(a^m) = (1 - q)q^{-m}ba^{m+1}.$$

This action has no classical meaning in geometry or algebraic geometry, because it would be trivial when $q = 1$ (the commutative case); it is our first example of a 'purely quantum phenomenon'. Nevertheless, if we work over $\mathbb{C}$ for example, the action of the rescaled generator $b/(q-1)$ as $q \to 1$ leaves a nonzero classical 'remnant' which can still be useful. [For example, the action of the special conformal transformations on classical $\mathbb{R}^n$ can be similarly be expressed as the remnant as $q \to 1$ of the adjoint action of a suitable q-deformed $\mathbb{R}_q^n$ on itself.]

Proposition 2.8 *(Left coregular action) If H' is dually paired with a bialgebra or Hopf algebra H, it acts on it by*

$$R_\phi^*(h) = \sum h_{(1)}\langle\phi, h_{(2)}\rangle$$

for all $\phi \in H'$, $h \in H$.

Proof It is easy to see that we have an action. It respects the product because $\phi\triangleright(hg) = (hg)_{(1)}\langle\phi, (hg)_{(2)}\rangle = h_{(1)}g_{(1)}\langle\phi, h_{(2)}g_{(2)}\rangle = (\phi_{(1)}\triangleright h)(\phi_{(2)}\triangleright g)$ and $\phi\triangleright 1 = 1\langle\phi, 1\rangle = \epsilon(\phi)$, for all $\phi \in H'$ and $h, g \in H$. $\qquad\square$

For our standard examples, we have (immediately from the definitions):

(i) for $k(G)$ acting on kG,

$$R_\phi^*(g) = \phi(g)g, \quad \forall g \in G, \ \phi \in k(G).$$

(ii) for $\mathbb{C}[G]$ acting on $U(\mathfrak{g})$,

$$R_{x^i{}_j}^*(\xi) = 1\rho(\xi)^i{}_j + \xi\delta^i{}_j, \quad \forall \xi \in \mathfrak{g}.$$

(iii) for kG acting on $k(G)$,

$$R_g^*(\phi)(h) = \phi(hg), \quad \forall h, g \in G, \ \phi \in k(G).$$

(iv) for $U(\mathfrak{g})$ acting on $\mathbb{C}[G]$,

$$R_\xi^*(x^i{}_j) = \sum_k x^i{}_k \rho(\xi)^k{}_j, \quad \forall \xi \in \mathfrak{g}.$$

(v) for $U_q(b_+)$ paired with itself,

$$R_g^*(X^m g^n) = q^n X^m g^n, \quad R_X^*(X^m g^n) = [m]_q X^{m-1} g^{n+1},$$

where we used the formula for ΔX^m stated in the last lecture, and the duality pairing above.

Of these, (iii) and (iv) are geometrical. They are respectively the action on functions by right translation, and the action of a Lie algebra on its coordinate ring as the associated left-invariant vector field. The example (v) has an aspect of this but in a q-deformed setting. The action of X involves the 'q-derivative'

$$\partial_q(X^m) = [m]_q X^{m-1}.$$

This is often written conveniently as

$$\partial_q f(X) = \frac{f(X) - f(qX)}{X(1 - q)}$$

for any polynomial $f(X)$ (the numerator here is always divisible by $X(1 - q)$).

3

Coactions. Quantum plane $\mathbb{A}_q^2$

In the last lecture we studied the notion of a bialgebra or Hopf algebra acting on an algebra. In keeping with our philosophy of treating both algebras and coalgebras democratically, we have clearly the notion:

Definition 3.1 *A coalgebra* (C, Δ, ϵ) *is an* H-module coalgebra *if*
1. *C is an H-module.*
2. *$\Delta : C \to C \otimes C$ and $\epsilon : C \to k$ commute with the action of H.*
Explicitly,

$$\Delta(h \triangleright c) = \sum h_{(1)} \triangleright c_{(1)} \otimes h_{(2)} \triangleright c_{(2)}, \quad \epsilon(h \triangleright c) = \epsilon(h)\epsilon(c), \quad \forall h \in H, \ c \in C.$$

This is obviously less familiar than the notion of module algebra, since coalgebras are less well known. There are, however, plenty of natural examples.

Proposition 3.2 *(Coadjoint action) If H' is dually paired with H, it acts on it as a coalgebra by*

$$\mathrm{Ad}_\phi^*(h) = \sum h_{(2)} \langle \phi, (Sh_{(1)})h_{(3)} \rangle, \quad \forall h \in H, \ \phi \in H'.$$

Proof This is an action since

$$
\begin{aligned}
\phi \triangleright (\psi \triangleright h) &= (\psi \triangleright h)_{(2)} \langle \phi, (S(\psi \triangleright h)_{(1)})(\psi \triangleright h)_{(3)} \rangle \\
&= h_{(3)} \langle \psi, (Sh_{(1)})h_{(5)} \rangle \langle \phi, (Sh_{(2)})h_{(4)} \rangle \\
&= h_{(3)} \langle \psi_{(1)}, Sh_{(1)} \rangle \langle \psi_{(2)}, h_{(5)} \rangle \langle \phi_{(1)}, Sh_{(2)} \rangle \langle \phi_{(2)}, h_{(4)} \rangle \\
&= h_{(2)} \langle \phi_{(1)} \psi_{(1)}, Sh_{(1)} \rangle \langle \phi_{(2)} \psi_{(2)}, h_{(3)} \rangle \\
&= h_{(2)} \langle \phi\psi, (Sh_{(1)})h_{(3)} \rangle = (\phi\psi) \triangleright h.
\end{aligned}
$$

We used the definitions Ad^*, the pairing axioms, Proposition 1.4 that S is an anticoalgebra map, and the pairing axioms again. Also $1 \triangleright h =$

17

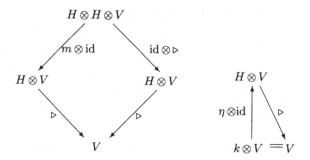

Fig. 3.1. Axioms of an action expressed as commutative diagrams.

$h_{(2)}\langle 1, (Sh_{(1)})h_{(3)}\rangle = h_{(2)}\epsilon((Sh_{(1)})h_{(3)}) = h.$ That the coproduct is respected is

$$
\begin{aligned}
\mathrm{Ad}^*_{\phi_{(1)}}(h_{(1)}) &\otimes \mathrm{Ad}^*_{\phi_{(2)}}(h_{(2)})\\
&= h_{(2)}\langle\phi_{(1)}, (Sh_{(1)})h_{(3)}\rangle \otimes h_{(5)}\langle\phi_{(2)}, (Sh_{(4)})h_{(6)}\rangle\\
&= h_{(2)} \otimes h_{(5)}\langle\phi, (Sh_{(1)})h_{(3)}(Sh_{(4)})h_{(6)}\rangle\\
&= h_{(2)} \otimes h_{(3)}\langle\phi, (Sh_{(1)})h_{(4)}\rangle = \Delta \circ \mathrm{Ad}^*_{\phi}(h),
\end{aligned}
$$

using the pairing axioms and the antipode axiom. That the counit is respected is immediate.   □

Next, continuing our philosophy of completing our ideas to include all their arrow-reversals, we should have the dual notion to that of an action itself. To arrive at this, one should write out the notion of action a bit more formally as commutative diagrams. This is shown in Figure 3.1 for an algebra acting on a vector space. Our other definitions can equally well be written as diagrams and their arrows reversed. For example, the notion of C being an H-module coalgebra (in addition to being an H-module) is shown in Figure 3.2. The arrow-reversal of Figure 3.1 is clearly the notion of a coaction:

Definition 3.3 *A* coaction *of a coalgebra C on a vector space V is a map $\beta : V \to C \otimes V$ such that*

  1. $(\mathrm{id} \otimes \beta) \circ \beta = (\Delta \otimes \mathrm{id})\beta.$
  2. $\mathrm{id} = (\epsilon \otimes \mathrm{id}) \circ \beta.$

Fig. 3.2. Additional axioms for an H-module coalgebra expressed as commutative diagrams.

We write

$$\beta(v) = \sum v^{(\bar{1})} \otimes v^{(\bar{2})}, \quad \forall v \in V$$

as an explicit notation, and say that V is a comodule.

By definition, a coaction of a Hopf algebra means as a coalgebra. So a bialgebra or Hopf algebra can either act as an algebra or coact as a coalgebra. Similarly, the arrow-reversal of the notion of a module coalgebra in Figure 3.2 is:

Definition 3.4 *A bialgebra or Hopf algebra H coacts on an algebra A (an H-comodule algebra) if*

1. A is an H-comodule.

2. The coaction $\beta : A \to H \otimes A$ is an algebra homomorphism, where $H \otimes A$ has the tensor product algebra structure.

Clearly, a coalgebra always coacts on itself by its coproduct (the regular coaction). This is the arrow-reversal of the notion of the regular representation of an algebra. In the bialgebra or Hopf algebra case, $\Delta : H \to H \otimes H$ clearly makes H into an H-comodule algebra. Similarly, we have the notion of an H-*comodule coalgebra* given by arrow-reversing the notion of module algebra. We leave this as an exercise.

Also, two comodules V, W have a tensor product

$$\beta_{V \otimes W}(v \otimes w) = \sum v^{(\bar{1})} w^{(\bar{1})} \otimes v^{(\bar{2})} \otimes w^{(\bar{2})}, \quad \forall v \in V, \ w \in W.$$

The trivial comodule is $V = k$ with

$$\beta(\lambda) = 1 \otimes \lambda, \quad \forall \lambda \in k.$$

A morphism or 'intertwiner' $\phi : V \to W$ between two comodules is a map commuting with the coaction of H in the sense

$$(\mathrm{id} \otimes \phi)\beta_V = \beta_W \circ \phi.$$

All this may look unfamiliar, but it is just the arrow-reversal of the tensor product etc., for actions in the last lecture. In these terms, the notion of H-comodule algebra says equivalently that the product $m : A \otimes A \to k$ and unit $\eta : k \to A$ commute with the coaction of H. Equivalently, an H-comodule coalgebra says that the coalgebra structure maps commute with the coaction of H.

Proposition 3.5 *Every Hopf algebra coacts on itself as a coalgebra, by*

$$\mathrm{Ad}(h) = \sum h_{(1)} S h_{(3)} \otimes h_{(2)}.$$

Proof More exercise in the subscript notation. That it is a coaction is

$$(\mathrm{id} \otimes \mathrm{Ad})\mathrm{Ad}(h) = h_{(1)} S h_{(3)} \otimes \mathrm{Ad}(h_{(2)}) = h_{(1)} S h_{(5)} \otimes h_{(2)} S h_{(4)} \otimes h_{(3)}$$
$$= h_{(1)(1)} (S h_{(3)})_{(1)} \otimes h_{(1)(2)} (S h_{(3)})_{(2)} \otimes h_{(2)} = (\Delta \otimes \mathrm{id})\mathrm{Ad}(h)$$

and $h_{(1)} S h_{(3)} \epsilon(h_{(2)}) = h_{(1)} S h_{(2)} = \epsilon(h)$. That it respects the coproduct is

$$\mathrm{Ad} \circ \Delta h = h_{(1)}{}^{(\bar{1})} h_{(2)}{}^{(\bar{1})} \otimes h_{(1)}{}^{(\bar{2})} \otimes h_{(2)}{}^{(\bar{2})} = h_{(1)} (S h_{(3)}) h_{(4)} S h_{(6)} \otimes h_{(2)} \otimes h_{(5)}$$
$$= (\mathrm{id} \otimes \Delta)\mathrm{Ad}(h)$$

using the tensor product coaction and the antipode axioms. That it respects the counit is trivial. ∎

Finally, as well as arrow-reversal symmetry, we have the usual left–right symmetry of constructions in linear algebra. We have a right-handed version of the construction of actions in the last lecture, and a right-handed version $\beta : V \to V \otimes H$ of the notion of coaction. For example, the coproduct $\Delta : H \to H \otimes H$ can equally well be viewed as a right coaction of H on itself as an algebra (the right regular coaction). Similarly, H has a right coaction on itself as a coalgebra by

$$\mathrm{Ad}_R(h) = h_{(2)} \otimes (S h_{(1)}) h_{(3)}, \quad \forall h \in H.$$

Moreover:

Lemma 3.6 *If H' is dually paired to H then a right coaction of H implies a left action of H' by evaluation. An H-comodule algebra becomes an H'-module algebra. An H-comodule coalgebra becomes an H'-module coalgebra.*

Proof Elementary – use the axioms of a duality pairing. ∎

Thus, the left coregular action in the last lecture is the dualisation of the right regular coaction given by the coproduct $\Delta : H \to H \otimes H$ viewed as a right coaction. (This is the reason we called it R^*.) And in the case of a finite-dimensional bialgebra or Hopf algebra, a coaction of H is fully equivalent to an action of H^*. In the converse direction, if $\triangleright$ is a left action of H^* then

$$\beta(v) = \sum_a f^a \triangleright v \otimes e_a, \quad \forall v \in V$$

is the corresponding right coaction of H, where $\{e_a\}$ is a basis of H and $\{f^a\}$ is its dual basis. In the infinite-dimensional case the notion of coaction is stricter than the notion of action: not every action of H' comes from a coaction of H, even when the pairing is nondegenerate. Supposing a coaction typically keeps us in an algebraic setting, and we will use them often for this reason.

Example 3.7 *(The quantum plane) The algebra* $\mathbb{A}_q^2$ *is* $k\langle x, y \rangle$ *modulo the relations* $yx = qxy$. *It is a right* $SL_q(2)$-*comodule algebra under*

$$\beta(x) = x \otimes a + y \otimes c, \quad \beta(y) = x \otimes b + y \otimes d.$$

Proof We check first that β as stated extends as an algebra map to products of the generators. Thus, $\beta(yx) = (x \otimes b + y \otimes d)(x \otimes a + y \otimes c) = x^2 \otimes ba + y^2 \otimes dc + yx \otimes da + xy \otimes bc = qx^2 \otimes ab + qy^2 \otimes cd + yx \otimes ad + yx \otimes (q - q^{-1})bc + xy \otimes bc = qx^2 \otimes ab + qy^2 \otimes cd + qxy \otimes ad + yx \otimes qcb = q(x \otimes a + y \otimes c)(x \otimes b + y \otimes d) = \beta(qxy)$ using the q-commutativity relations of $SL_q(2)$, followed by the relations of $\mathbb{A}_q^2$. Hence, the map $\beta : \mathbb{A}_q^2 \to \mathbb{A}_q^2 \otimes SL_q(2)$ is well-defined as an algebra map. Next, it is trivial to see that β on the generators obeys the comodule axiom with respect to the matrix form of the coproduct of $SL_q(2)$. Hence, by induction, this extends to all products. A basis of $\mathbb{A}_q^2$ here is clearly of the form $\{x^n y^m\}$ for $n, m \in \mathbb{Z}_+$. $\qquad\square$

A useful shorthand for the coaction in this example is

$$\beta(x, y) = (x, y) \otimes \begin{pmatrix} a & b \\ c & d \end{pmatrix},$$

where the matrix action on the covector (x, y) is to be understood. This coaction is a q-deformed analogue, in our algebraic terms, of the classical action of SL_2 on k^2. This is clearly the beginning of 'quantum linear algebra'. Similarly, there is a left coaction

$$\beta_L \begin{pmatrix} x \\ y \end{pmatrix} = \begin{pmatrix} a & b \\ c & d \end{pmatrix} \otimes \begin{pmatrix} x \\ y \end{pmatrix},$$

which is a notation for

$$\beta_L(x) = a \otimes x + b \otimes y, \quad \beta_L(y) = c \otimes x + d \otimes y.$$

Note that we did not actually use the q-determinant relation in the above proof; both of these coactions on the quantum plane lift to coactions of the bialgebra $M_q(2)$ defined as $k\langle a, b, c, d\rangle$ modulo the six q-commutativity relations given in the last lecture for $SL_q(2)$, without imposing the additional q-determinant relation. This is called the bi-algebra of 2×2 'quantum matrices'. In passing, we also note that the element $D = ad - q^{-1}bc$, which is grouplike and central in $M_q(2)$, can be formally inverted by adjoining D^{-1} to the algebra. This is the quantum group $GL_q(2) = M_q(2)[D^{-1}]$, where D^{-1} is added as an additional central generator with relations $DD^{-1} = D^{-1}D$. The coproduct is extended as

$$\Delta D^{-1} = D^{-1} \otimes D^{-1}, \quad \epsilon(D^{-1}) = 1$$

and the antipode is

$$S \begin{pmatrix} a & b \\ c & d \end{pmatrix} = D^{-1} \begin{pmatrix} d & -qb \\ -q^{-1}c & a \end{pmatrix}$$

in our matrix notation.

4

Automorphism quantum groups

In this lecture we will see that there are, as the astronomer Carl Sagan used to say, 'billions upon billions' of truly quantum groups, or at least noncommutative and noncocommutative bialgebras. Just as every reasonable space has an associated 'diffeomorphism group', so every finite-dimensional algebra has a 'diffeomorphism or automorphism' quantum group or *comeasuring bialgebra* associated to it. We use the latter technical term to avoid confusion with the usual automorphism group, which also exists, but which is too restrictive to serve in the correct geometrical role (of diffeomorphisms) in noncommutative geometry.

Definition 4.1 *Let A be an algebra. A comeasuring of A is a pair (B, β) where*

1. *B is an algebra.*
2. *$\beta : A \to A \otimes B$ is an algebra map.*

This is like a right comodule algebra but we only require B to be an algebra and hence do not require the comodule property itself. Morphisms between comeasurings are, by definition, maps between their underlying algebras connecting the corresponding β. We define $(M(A), \beta_U)$, when it exists, to be the initial universal object in the category of comeasurings of A, i.e. a comeasuring such that for any other comeasuring (B, β), there exists a unique algebra map $\pi : M(A) \to B$ such that $\beta = (\mathrm{id} \otimes \pi) \circ \beta_U$. Like all universal objects, if it exists it is unique up to unique isomorphism.

Proposition 4.2 *Let A be an algebra. Then $M(A)$, if it exists, is a bialgebra and β_U makes A an $M(A)$-comodule algebra. Any other coaction of a bialgebra on A as an algebra is the push-out of this one.*

Proof We note that $(M(A) \otimes M(A), (\beta_U \otimes \mathrm{id}) \circ \beta_U)$ is also a comeasuring. Hence there is an algebra map $\Delta : M(A) \to M(A) \otimes M(A)$ and $(\beta_U \otimes \mathrm{id}) \circ \beta_U = (\mathrm{id} \otimes \Delta) \circ \beta_U$. It remains to show that Δ is coassociative. For this, consider $(M(A)^{\otimes 3}, (\beta_U \otimes \mathrm{id} \otimes \mathrm{id}) \circ (\beta_U \otimes \mathrm{id}) \circ \beta_U)$ as another comeasuring. This induces an algebra map $\Delta_2 : M(A) \to M(A)^{\otimes 3}$ such that $(\mathrm{id} \otimes \Delta_2) \circ \beta_U$ is the comeasuring map associated to $M(A)^{\otimes 3}$. Both $(\Delta \otimes \mathrm{id}) \circ \Delta$ and $(\mathrm{id} \otimes \Delta) \circ \Delta$ clearly fulfil the role of Δ_2, and since the induced map is unique, they coincide with Δ_2 and hence with each other. Here,

$$
\begin{aligned}
(\beta_U \otimes \mathrm{id} \otimes \mathrm{id}) \circ (\beta_U \otimes \mathrm{id}) \circ \beta_U &= (\beta_U \otimes \mathrm{id} \otimes \mathrm{id}) \circ (\mathrm{id} \otimes \Delta) \circ \beta_U \\
&= (\mathrm{id} \otimes (\mathrm{id} \otimes \Delta) \circ \Delta) \circ \beta_U \\
(\beta_U \otimes \mathrm{id} \otimes \mathrm{id}) \circ (\beta_U \otimes \mathrm{id}) \circ \beta_U &= (\mathrm{id} \otimes \Delta \otimes \mathrm{id}) \circ (\beta_U \otimes \mathrm{id}) \circ \beta_U \\
&= (\mathrm{id} \otimes (\Delta \otimes \mathrm{id}) \circ \Delta) \circ \beta_U
\end{aligned}
$$

using $(\beta_U \otimes \mathrm{id}) \circ \beta_U = (\mathrm{id} \otimes \Delta) \circ \beta_U$ twice in each case. Moreover, $(k, \beta_k = \mathrm{id} \otimes 1)$ is a comeasuring and $\epsilon : M(A) \to k$ is the induced map. It is easy to see that it provides a counit. Finally, given any other coaction β of a bialgebra B on A as an algebra (i.e. if A is a B-comodule algebra), the fact that (B, β) is a comeasuring induces an algebra map $\pi : M(A) \to B$ and the coaction of B has the form $(\mathrm{id} \otimes \pi) \circ \beta_U$. Now consider $(B \otimes B, (\mathrm{id} \otimes \Delta) \circ \beta)$ as a comeasuring. We can also write this as $(B \otimes B, (\beta \otimes \mathrm{id}) \circ \beta)$ since β is a coaction. The universal property of $M(A)$ therefore induces a map $M(A) \to B \otimes B$ which we identify as either $\Delta \circ \pi$ or $(\pi \otimes \pi) \circ \Delta$, i.e. these coincide. Here

$$
(\mathrm{id} \otimes \Delta) \circ \beta = (\mathrm{id} \otimes \Delta \circ \pi) \circ \beta_U
$$

$$
(\beta \otimes \mathrm{id}) \circ \beta = (\mathrm{id} \otimes \pi \otimes \pi) \circ (\beta_U \otimes \mathrm{id}) \circ \beta_U = (\mathrm{id} \otimes (\pi \otimes \pi) \circ \Delta) \circ \beta_U.
$$

Similarly, $(k, (\mathrm{id} \otimes \epsilon) \circ \beta)$ is a comeasuring and induces the map $\epsilon \circ \pi$, but also coincides with the comeasuring (k, β_k) which corresponds to $\epsilon : M(A) \to k$. Hence $\pi : M(A) \to B$ is a bialgebra map and the coaction of B is induced from β_U by composing with π. $\qquad \square$

By a bialgebra map in this proof, we mean of course a map between bialgebras respecting both the algebra and coalgebra structures. Meanwhile, the term 'push-out' is the arrow-reversal of the usual notion of 'pull-back' of modules under an algebra map: if $\pi : C \to D$ is a coalgebra map between two coalgebras, and if C coacts on something by a coaction β, then $(\mathrm{id} \otimes \pi) \circ \beta$ is a coaction of D on the same object, called the *push-out* coaction.

We will now prove the existence of $M(A), \beta_U$ in the case where A is finite dimensional, and give an explicit description of it (in the infinite-dimensional case one generally needs a topological completion of some kind; alternatively one can reverse all arrows and define a measuring bialgebra which is fully algebraic but harder to work with for other reasons).

In fact, it is convenient to prove existence first in the nonunital setting. (Until now we have required that all our algebras have a unit and that algebra maps respect them – we will continue to assume this unless explicitly stated otherwise.) On the other hand, if A is not necessarily unital we can still follow the same ideas and define a measuring as (B, β) where B is a not necessarily unital algebra and β is merely multiplicative. Morphisms and the universal object are likewise defined with π a multiplicative map, i.e. we drop all axioms relating to the unit. In this case it is clear that the universal object $M(A)$ is a not necessarily unital bialgebra. We can, however, always formally adjoin a unit to it. We define $M_1(A) = M(A) \oplus k$ with $1.1 = 1$ and $a.1 = 1.a = a$ for $1 \in k$ and $a \in M(A)$ (the usual way to adjoin a unit to a nonunital algebra), and extend Δ, ϵ by $\Delta(1) = 1 \otimes 1$ and $\epsilon(1) = 1$ to yield a usual bialgebra, with unit.

Proposition 4.3 *Let A be a finite-dimensional not necessarily unital algebra. Then $M_1(A)$ exists and has the following form. Fix a basis $\{e_i \mid i = 0, \ldots, \dim(A) - 1\}$ of A. Then $M_1(A) = k\langle t^i{}_j \rangle$ (the free associative algebra on a matrix of $\dim(A)^2$ generators) modulo the relations*

$$\sum_a c_{ij}{}^a t^k{}_a = \sum_{a,b} c_{ab}{}^k t^a{}_i t^b{}_j, \quad \forall i, j, k,$$

where $c_{ij}{}^k$ are the structure constants of A defined by $e_i e_j = \sum_k c_{ij}{}^k e_k$. The coproduct and coaction are

$$\Delta t^i{}_j = \sum_a t^i{}_a \otimes t^a{}_j, \quad \epsilon(t^i{}_j) = \delta^i{}_j, \quad \beta_U(e_i) = \sum_a e_a \otimes t^a{}_i.$$

Proof Removing the unit, the corresponding $M(A)$ here is generated by $t^i{}_j$ without unit, with the relations shown. By construction it is a not necessarily unital algebra. The stated β_U is multiplicative since

$$\beta_U(e_i e_j) = \sum_k c_{ij}{}^k \beta_U(e_k) = \sum_{a,k} c_{ij}{}^a e_k \otimes t^k{}_a = \sum_{a,b,k} c_{ab}{}^k e_k \otimes t^a{}_i t^b{}_j$$

$$= \sum_{a,b} (e_a \otimes t^a{}_i)(e_b \otimes t^b{}_j) = \beta_U(e_i)\beta_U(e_j).$$

So we have a (nonunital) comeasuring. Now let (B, β) be another, and define $\bar{t}^i{}_j \in B$ by $\beta(e_i) = \sum_a e_a \otimes \bar{t}^a{}_i$. That β is multiplicative is the assertion that $\sum_a c_{ij}{}^a \bar{t}^k{}_a = \sum_{a,b} c_{ab}{}^k \bar{t}^a{}_i \bar{t}^b{}_j$, by the same computation as just made (we do not write out both). Hence $\pi : M(A) \to B$ defined by $\pi(t^i{}_j) = \bar{t}^i{}_j$ extends as a multiplicative map. Thus $M(A)$ has the required universal property. When we adjoin 1, we obtain the bialgebra $M_1(A)$ as stated. $\qquad\square$

The coproduct here has the same matrix form that we have seen for $k[G]$, $SL_q(2)$ and the 'quantum matrices' $M_q(2)$. The coaction has the same linear transformation form that we have seen on the generators of the quantum plane under the coaction $\beta : \mathbb{A}_q^2 \to \mathbb{A}_q^2 \otimes M_q(2)$ in the last lecture.

Now suppose that A is unital and that $e_0 = 1$ is a basis element. We let e_i, $i = 1, \ldots, \dim(A) - 1$ be the remaining basis elements.

Corollary 4.4 *The unital comeasuring bialgebra $M(A)$ is the quotient of $M_1(A)$ by the (ideal generated by the) relations $t^0{}_0 = 1$, $t^i{}_0 = 0$ for $i = 1, \ldots, \dim(A) - 1$.*

Proof Since $M(A)$ is a quotient of $M_1(A)$, it remains an algebra by definition and we inherit a map $\beta_U : A \to A \otimes M(A)$ given by the same formula as before. Since $t^0{}_0 = 1$ and $t^i{}_0 = 0$, we have now $\beta_U(1) = e_0 \otimes t^0{}_0 = 1 \otimes 1$ as required. The universal property holds for unital (B, β) since these obey $\beta(1) = 1 \otimes 1$ as well. $\qquad\square$

Note that it is also possible to work directly with $M(A)$ as $k\langle b_i, t^i{}_j \rangle$ modulo some (more complicated) relations and with coalgebra and coaction

$$\Delta t^i{}_j = \sum_a t^i{}_a \otimes t^a{}_j, \quad \Delta b_i = \sum_a b_a \otimes t^a{}_i + 1 \otimes b_i, \quad \epsilon(t^i{}_j) = \delta^i{}_j,$$

$$\epsilon(b_i) = 0, \quad \beta_U(e_i) = 1 \otimes b_i + \sum_a e_a \otimes t^a{}_i,$$

where the indices run $1, \ldots, \dim(A) - 1$. Here $b_i \equiv t^0{}_i$ and $t^i{}_j$ are the generators remaining in the quotient of $M_1(A)$, and this coalgebra and coaction are the inherited ones from the matrix form of coalgebra and coaction of that.

Also note that up to isomorphism, $M_1(A)$ is basis-independent and $M(A)$ is independent of the choice of the decomposition $A = k \oplus A'$ and basis of A', where k is spanned by 1 and A' is a complement. This follows from our abstract definition of these objects.

Example 4.5 *Let $A = k[i]/\langle i^2 + 1 \rangle$. Then $M_1(A)$ is $k\langle a, b, c, d \rangle$ modulo the relations*

$$a^2 - c^2 = a = d^2 - b^2, \quad ac + ca = c = -(bd + db),$$

$$ab - cd = b = ba - dc, \quad ad + cb = d = bc + da$$

and has the matrix form of coalgebra

$$\Delta \begin{pmatrix} a & b \\ c & d \end{pmatrix} = \begin{pmatrix} a & b \\ c & d \end{pmatrix} \otimes \begin{pmatrix} a & b \\ c & d \end{pmatrix}, \quad \epsilon \begin{pmatrix} a & b \\ c & d \end{pmatrix} = \begin{pmatrix} 1 & 0 \\ 0 & 1 \end{pmatrix}.$$

The unital comeasuring bialgebra $M(A)$ is $k\langle b, d \rangle$ modulo the relations

$$d^2 - b^2 = 1, \quad bd + db = 0$$

and has the coalgebra

$$\Delta d = d \otimes d, \quad \Delta b = b \otimes d + 1 \otimes b, \quad \epsilon(d) = 1, \quad \epsilon(b) = 0.$$

Proof We choose the basis $e_0 = 1$ and $e_1 = i$. Then

$$c_{00}{}^0 = 1, \quad c_{01}{}^1 = 1, \quad c_{10}{}^1 = 1, \quad c_{11}{}^0 = -1$$

and all others are zero. We put this into the explicit formulae for $M_1(A)$; four of the eight relations are redundant while the remainder come out as shown. For example

$$c_{11}{}^0 t^1{}_0 = \sum_{a,b} c_{ab}{}^1 t^a{}_1 t^b{}_1 = t^0{}_1 t^1{}_1 + t^1{}_1 t^0{}_1,$$

which is the relation $-c = bd + db$. We then set $a = 1$ and $c = 0$ to obtain $M(A)$. $\qquad \square$

Note that $M_1(A)$ here has another interesting quotient, namely by the relations $c = -b$ and $d = a$. This is $k\langle a, b \rangle$ modulo the relations

$$ab + ba = b, \quad a^2 - b^2 = a$$

and the coalgebra

$$\Delta a = a \otimes a - b \otimes b, \quad \Delta b = b \otimes a + a \otimes b, \quad \epsilon(a) = 1, \quad \epsilon(b) = 0.$$

This is called the 'trigonometric bialgebra' due to the similarity here with the addition rule

$$\cos(x + y) = \cos(x)\cos(y) - \sin(x)\sin(y),$$

$$\sin(x + y) = \sin(x)\cos(y) + \cos(x)\sin(y)$$

for cos and sin. The counit corresponds to $\cos(0) = 1$ and $\sin(0) = 0$. And when $\imath = \sqrt{-1}$ exists in k, we may write the relations and coalgebra here as

$$(a \pm \imath b)^2 = a \pm \imath b, \quad \Delta(a \pm \imath b) = (a \pm \imath b) \otimes (a \pm \imath b), \quad \epsilon(a \pm \imath b) = 1,$$

which is analogous to $\exp(x) = \cos(x) + \imath\sin(x)$.

When $k = \mathbb{R}$, we have computed here the 'automorphism quantum group' of $\mathbb{C}$ as a 2-dimensional algebra over $\mathbb{R}$. In general, if $m(x)$ is a monic irreducible polynomial then $A = k[x]/\langle m(x)\rangle$ is a field extension of k (which is embedded via 1), and $M(A)$ can be viewed as a natural 'Galois quantum group' associated to it. The role of such objects in number theory, however, is totally unexplored at the moment.

5

Quasitriangular structures

In this lecture we start what can be called 'Drinfeld theory'. We have seen that the classical objects kG and $U(\mathfrak{g})$ are cocommutative, i.e. $\tau \circ \Delta = \Delta$, while their duals $k(G)$, $k[G]$ etc. are commutative, whereas a true quantum group should be neither. But we should not drop (co)commutativity altogether, otherwise we will have a theory with properties much weaker than group theory or Lie theory. The idea of Drinfeld theory is to drop cocommutativity but keep it under control by means of a 'cocycle' of some kind.

Definition 5.1 *A* quasitriangular *bialgebra or Hopf algebra is a pair* $(H, \mathcal{R})$, *where*

1. *H is a bialgebra or Hopf algebra.*
2. *$\mathcal{R} \in H \otimes H$ is invertible and obeys*

$$\tau \circ \Delta h = \mathcal{R}(\Delta h)\mathcal{R}^{-1}, \ \forall h \in H,$$

where the product is in $H \otimes H$.

3.

$$(\Delta \otimes \mathrm{id})\mathcal{R} = \mathcal{R}_{13}\mathcal{R}_{23}, \quad (\mathrm{id} \otimes \Delta)\mathcal{R} = \mathcal{R}_{13}\mathcal{R}_{12},$$

where the product is in $H \otimes H \otimes H$ and $\mathcal{R}_{12} \equiv \mathcal{R} \otimes 1$, $\mathcal{R}_{23} \equiv 1 \otimes \mathcal{R}$ as elements of this. The numerical suffixes denote the embedding of $\mathcal{R}$ in higher tensor powers of H.

The specific meaning of such a quasitriangular structure $\mathcal{R}$ will emerge during the next several lectures. However, there are many examples and indeed, most self-respecting quantum groups are either quasitriangular or dual to a quasitriangular one.

Lemma 5.2 *Let $(H, \mathcal{R})$ be a quasitriangular bialgebra. Then*

1. $(\epsilon \otimes \mathrm{id})\mathcal{R} = (\mathrm{id} \otimes \epsilon)\mathcal{R} = 1$.

2. $(H, \mathcal{R}_{21}^{-1})$ *is also a quasitriangular bialgebra* $(\mathcal{R}_{21}^{-1} = \tau(\mathcal{R}^{-1})$ *is called the 'conjugate' quasitriangular structure)*.

3. $\mathcal{R}_{12}\mathcal{R}_{13}\mathcal{R}_{23} = \mathcal{R}_{23}\mathcal{R}_{13}\mathcal{R}_{12}$ *holds in* $H \otimes H \otimes H$ *(the Yang–Baxter equation)*.

Proof For the first part, apply ϵ to the third axiom of a quasitriangular structure, thus $(\epsilon \otimes \mathrm{id} \otimes \mathrm{id})(\Delta \otimes \mathrm{id})\mathcal{R} = \mathcal{R}_{23} = (\epsilon \otimes \mathrm{id} \otimes \mathrm{id})\mathcal{R}_{13}\mathcal{R}_{23}$, so that $(\epsilon \otimes \mathrm{id})\mathcal{R} = 1$ because $\mathcal{R}_{23}$ is invertible. Similarly for the other side. That $\mathcal{R}_{21}^{-1}$ is another quasitriangular structure is an elementary exercise from the definitions. For example, apply τ to the second axiom to obtain $\Delta h = \mathcal{R}_{21}(\tau \circ \Delta h)\mathcal{R}_{21}^{-1}$, i.e. $\mathcal{R}_{21}^{-1}(\Delta h)\mathcal{R}_{21} = \tau \circ \Delta h$, which is the same axiom for $\mathcal{R}_{21}^{-1}$ in the role of $\mathcal{R}$. For the last part of the lemma we compute $(\mathrm{id} \otimes \tau \circ \Delta)\mathcal{R}$ in two ways: using the third axiom directly, or using the second axiom and then the third axiom. Thus, $(\mathrm{id} \otimes \tau \circ \Delta)\mathcal{R} = (\mathrm{id} \otimes \tau)(\mathrm{id} \otimes \Delta)\mathcal{R} = (\mathrm{id} \otimes \tau)\mathcal{R}_{13}\mathcal{R}_{12} = \mathcal{R}_{12}\mathcal{R}_{13}$ and $(\mathrm{id} \otimes \tau \circ \Delta)\mathcal{R} = \mathcal{R}_{23}((\mathrm{id} \otimes \Delta)\mathcal{R})\mathcal{R}_{23}^{-1} = \mathcal{R}_{23}\mathcal{R}_{13}\mathcal{R}_{12}\mathcal{R}_{23}^{-1}$. $\square$

The Yang–Baxter equation arises in physics and also in knot theory as we will see later in the course. In either case one looks for matrix solutions of it; we see that for every finite-dimensional representation of H, the image of $\mathcal{R}$ provides such a matrix [so it is sometimes called the 'universal R-matrix' for this reason].

Lemma 5.3 *Let* $(H, \mathcal{R})$ *be a quasitriangular Hopf algebra. Then*

$$(S \otimes \mathrm{id})\mathcal{R} = \mathcal{R}^{-1}, \qquad (\mathrm{id} \otimes S)\mathcal{R}^{-1} = \mathcal{R},$$

and hence $(S \otimes S)\mathcal{R} = \mathcal{R}$ *(i.e.* $\mathcal{R}$ *is* S-*invariant).*

Proof We write $\mathcal{R} \equiv \sum \mathcal{R}^{(1)} \otimes \mathcal{R}^{(2)}$. Then $\mathcal{R}^{(1)}{}_{(1)} S\mathcal{R}^{(1)}{}_{(2)} \otimes \mathcal{R}^{(2)} = 1$ by the antipode axiom and the preceding lemma, but also equals $\mathcal{R}(S \otimes \mathrm{id})\mathcal{R}$ by axiom 3 of a quasitriangular structure. Similarly for the other side; hence, $(S \otimes \mathrm{id})\mathcal{R} = \mathcal{R}^{-1}$. Similarly for $(\mathrm{id} \otimes S)\mathcal{R}^{-1} = \mathcal{R}$ once we know that $(\Delta \otimes \mathrm{id})(\mathcal{R}^{-1}) = (\mathcal{R}_{13}\mathcal{R}_{23})^{-1} = \mathcal{R}_{23}^{-1}\mathcal{R}_{13}^{-1}$, etc. which follows since Δ is an algebra homomorphism. $\square$

Before proceeding with the general theory, let us see what a quasitriangular structure means for one of our simple classes of Hopf algebras.

Example 5.4 *A quasitriangular structure on a finite-group function algebra* $k(G)$ *requires precisely*

1. G is Abelian.

2. $\mathcal{R} \in k(G \times G)$ is a bicharacter.

Proof We identify $k(G) \otimes k(G) = k(G \times G)$, i.e. we need $\mathcal{R}$ a function of two variables on the group. It should be pointwise invertible (i.e. nowhere vanishing). Since the algebra is commutative, axiom 2 of a quasitriangular structure requires that the coproduct is cocommutative, i.e. G is Abelian. The remaining axiom 3 is clearly

$$\mathcal{R}(xy, z) = \mathcal{R}(x, z)\mathcal{R}(y, z), \quad \mathcal{R}(x, yz) = \mathcal{R}(x, z)\mathcal{R}(x, y), \quad x, y, z \in G.$$

The invertibility of $\mathcal{R}$ is equivalent to the counity property in the first lemma, i.e. $\mathcal{R}(x, e) = 1 = \mathcal{R}(e, x)$ for all $x \in G$ (here e is the group identity). These equations are precisely the definition of a bicharacter on G. $\qquad\square$

Actually, our present motivation comes not from this example $k(G)$ but from generalising the group algebra kG and Lie theory.

Definition 5.5 *A quasitriangular Hopf algebra $(H, \mathcal{R})$ is called*

1. Triangular if $Q \equiv \mathcal{R}_{21}\mathcal{R} \in H \otimes H$ is trivial in the sense $Q = 1 \otimes 1$.

2. Factorisable if Q is nondegenerate in the sense that the induced linear map $H^* \to H$ sending $\phi \mapsto (\phi \otimes \mathrm{id})(Q)$ is surjective.

Triangular Hopf algebras are the ones most similar to the cocommutative Hopf algebras kG, $U(\mathfrak{g})$, which can be trivially regarded as quasitriangular with $Q = \mathcal{R} = 1 \otimes 1$. More generally, the condition says that the quasitriangular structure and its conjugate coincide. Factorisable Hopf algebras are at the opposite extreme and are motivated in a different way from the theory of finite-dimensional complex semisimple Lie algebras, where the Killing form can be viewed as an isomorphism $\mathfrak{g} \to \mathfrak{g}^*$. For this reason the element Q is sometimes called the 'inverse quantum Killing form'. It has the corresponding Ad-invariance:

Proposition 5.6 *For any quasitriangular Hopf algebra H,*

$$\mathrm{Ad}_h((S \otimes \mathrm{id})(Q)) = \epsilon(h)(S \otimes \mathrm{id})(Q), \quad \forall h \in H,$$

where Ad *is the quantum group adjoint action extended to $H \otimes H$. This says that $(S \otimes \mathrm{id})(Q)$ is Ad-invariant.*

Proof We write $\sum \mathcal{R}'^{(1)} \otimes \mathcal{R}'^{(2)}$ as an explicit notation for a second copy of $\mathcal{R}$, with the prime used to distinguish it from the first copy. Then

$$
\begin{aligned}
\mathrm{Ad}_h((S \otimes \mathrm{id})(Q)) &= h_{(1)}(SQ^{(1)})Sh_{(2)} \otimes h_{(3)}Q^{(2)}Sh_{(4)} \\
&= h_{(1)}(S\mathcal{R}^{(1)})(S\mathcal{R}'^{(2)})Sh_{(2)} \otimes h_{(3)}\mathcal{R}'^{(1)}\mathcal{R}^{(2)}Sh_{(4)} \\
&= h_{(1)}(S\mathcal{R}^{(1)})Sh_{(3)}(S\mathcal{R}'^{(2)}) \otimes \mathcal{R}'^{(1)}h_{(2)}\mathcal{R}^{(2)}Sh_{(4)} \\
&= h_{(1)}Sh_{(2)}(S\mathcal{R}^{(1)})(S\mathcal{R}'^{(2)}) \otimes \mathcal{R}'^{(1)}\mathcal{R}^{(2)}h_{(3)}Sh_{(4)} \\
&= \epsilon(h)(S \otimes \mathrm{id})(Q),
\end{aligned}
$$

using that S is an antialgebra map, and axiom 2 of a quasitriangular structure for the third and fourth equalities. $\qquad \square$

There are many further interesting objects that can be built from $\mathcal{R}$. Using such objects, one can prove various results which are analogous to those for cocommutative Hopf algebras, but differing by something depending on $\mathcal{R}$ (since this controls the breakdown of cocommutativity). For example, we mentioned in Lecture 1 that $S^2 = \mathrm{id}$ for a commutative or cocommutative Hopf algebra, so in the quasitriangular case we can expect this to hold up to conjugation by something built from $\mathcal{R}$. In particular, this ensures that S is invertible for a quasitriangular Hopf algebra (although S is motivated by the idea of 'group inversion' we have not actually assumed that it is invertible for a general Hopf algebra, and indeed it need not be when the Hopf algebra is noncommutative and noncocommutative). The computations in the following theorem are 'as hard as it gets' and probably beyond what could be expected from the reader at this point: they are included mainly for completeness. Also, in 10 years no-one has ever found a better notation for this kind of computation, but the reader is welcome to try to think one up.

Theorem 5.7 *Let $(H, \mathcal{R})$ be a quasitriangular Hopf algebra. Then*

    1. S is invertible.

    2. There exists an invertible element $\mathrm{u} \in H$ such that $S^2(h) = \mathrm{u}h\mathrm{u}^{-1}$ for all $h \in H$ and $\Delta\mathrm{u} = Q^{-1}(\mathrm{u} \otimes \mathrm{u})$.

Proof (i) We define $\mathrm{u} = \sum(S\mathcal{R}^{(2)})\mathcal{R}^{(1)}$ and compute $(Sh_{(2)})\mathrm{u}h_{(1)} = (Sh_{(2)})(S\mathcal{R}^{(2)})\mathcal{R}^{(1)}h_{(1)} = (S(\mathcal{R}^{(2)}h_{(2)}))\mathcal{R}^{(1)}h_{(1)} = (S(h_{(1)}\mathcal{R}^{(2)}))h_{(2)}\mathcal{R}^{(1)} = (S\mathcal{R}^{(2)})(Sh_{(1)})h_{(2)}\mathcal{R}^{(1)} = \epsilon(h)\mathrm{u}$ for all $h \in H$, by the antipode properties, axiom 2 of a quasitriangular structure and the counit property. Then

$$
(S^2h)\mathrm{u} = (S^2h_{(2)})\epsilon(h_{(1)})\mathrm{u} = (S^2h_{(3)})(Sh_{(2)})\mathrm{u}h_{(1)} = \mathrm{u}h, \quad \forall h \in H.
$$

(ii) Next, we define $u^{-1} = \sum \mathcal{R}^{(2)} S^2 \mathcal{R}^{(1)}$ and verify that indeed $u^{-1}u =$ $\mathcal{R}^{(2)}(S^2\mathcal{R}^{(1)})u = \mathcal{R}^{(2)}u\mathcal{R}^{(1)} = \mathcal{R}^{(2)}(S\mathcal{R}'^{(2)})\mathcal{R}'^{(1)}\mathcal{R}^{(1)} = (S\mathcal{R}^{(2)})(S\mathcal{R}'^{(2)})$ $\mathcal{R}'^{(1)}S\mathcal{R}^{(1)} = S(\mathcal{R}'^{(2)}\mathcal{R}^{-(2)})\mathcal{R}'^{(1)}\mathcal{R}^{-(1)} = S(1)1 = 1$, using part (i) and the facts already established in Lemma 5.3 for the action of S on $\mathcal{R}$. Likewise, $uu^{-1} = u\mathcal{R}^{(2)}S^2\mathcal{R}^{(1)} = (S^2\mathcal{R}^{(2)})uS^2\mathcal{R}^{(1)} = (S\mathcal{R}^{(2)})uS\mathcal{R}^{(1)} =$ $\mathcal{R}^{(2)}u\mathcal{R}^{(1)} = 1$.

(iii) Next, we define $S^{-1}(h) = u^{-1}(Sh)u$ and verify that for all $h \in$ H we have $(S^{-1}h_{(2)})h_{(1)} = u^{-1}(Sh_{(2)})uh_{(1)} = u^{-1}(Sh_{(2)})(S^2h_{(1)})u =$ $u^{-1}(S((Sh_{(1)})h_{(2)}))u = \epsilon(h)u^{-1}u = \epsilon(h)$, using parts (i) and (ii) and the properties of S. Likewise, we verify that $h_{(2)}S^{-1}h_{(1)} = h_{(2)}u^{-1}(Sh_{(1)})u =$ $u^{-1}(S^2h_{(2)})(Sh_{(1)})u = u^{-1}(S(h_{(1)}Sh_{(2)}))u = \epsilon(h)u^{-1}u = \epsilon(h)$. From this it follows that

$$SS^{-1}h = (SS^{-1}h_{(1)})(Sh_{(2)})h_{(3)} = (S(h_{(2)}S^{-1}h_{(1)}))h_{(3)} = S(1)h = h,$$

$$S^{-1}Sh = (S^{-1}Sh_{(3)})(S^{-1}h_{(2)})h_{(1)} = (S^{-1}(h_{(2)}Sh_{(3)}))h_{(1)} = S^{-1}(1)h = h$$

for all $h \in H$, using antimultiplicativity of S (and hence of S^{-1}).

(iv) Finally, we compute

$$\begin{aligned}
\Delta u &= \Delta(S\mathcal{R}^{(2)})\mathcal{R}^{(1)} = (S\mathcal{R}^{(2)}{}_{(2)})\mathcal{R}^{(1)}{}_{(1)} \otimes (S\mathcal{R}^{(2)}{}_{(1)})\mathcal{R}^{(1)}{}_{(2)} \\
&= (S(\mathcal{R}^{(2)}{}_{(2)}\mathcal{R}'^{(2)}{}_{(2)}))\mathcal{R}^{(1)} \otimes (S(\mathcal{R}^{(2)}{}_{(1)}\mathcal{R}'^{(2)}{}_{(1)}))\mathcal{R}'^{(1)} \\
&= (S(\mathcal{R}''^{(2)}\mathcal{R}'''^{(2)}))\mathcal{R}''^{(1)}\mathcal{R}^{(1)} \otimes (S(\mathcal{R}^{(2)}\mathcal{R}'^{(2)}))\mathcal{R}'''^{(1)}\mathcal{R}'^{(1)} \\
&= (S\mathcal{R}'''^{(2)})u\mathcal{R}^{(1)} \otimes (S(\mathcal{R}^{(2)}\mathcal{R}'^{(2)}))\mathcal{R}'''^{(1)}\mathcal{R}'^{(1)} \\
&= (S\mathcal{R}'''^{(2)})\mathcal{R}^{-(1)}u \otimes (S\mathcal{R}'^{(2)})\mathcal{R}^{-(2)}\mathcal{R}'''^{(1)}\mathcal{R}'^{(1)} \\
&= \mathcal{R}^{-(1)}(S\mathcal{R}'''^{(2)})u \otimes \mathcal{R}^{-(2)}(S\mathcal{R}'^{(2)})\mathcal{R}'^{(1)}\mathcal{R}'''^{(1)} \\
&= \mathcal{R}^{-(1)}(S\mathcal{R}'''^{(2)})u \otimes \mathcal{R}^{-(2)}u\mathcal{R}'''^{(1)} \\
&= \mathcal{R}^{-(1)}(S\mathcal{R}'''^{(2)})u \otimes \mathcal{R}^{-(2)}(S^2\mathcal{R}'''^{(1)})u \\
&= \mathcal{R}^{-1}\mathcal{R}_{21}^{-1}(u \otimes u).
\end{aligned}$$

We freely use the antimultiplicativity and anticomultiplicativity properties of the antipode, starting with the latter. The third and fourth equalities use quasitriangularity axiom 3, and the fifth equality uses the facts about u in part (i) and the action of S on $\mathcal{R}$ in Lemma 5.3. The sixth equality is the Yang–Baxter equation applied in a suitable form. We then recognise another copy of u and use part (i) again. $\square$

6

Roots of unity. $u_q(sl_2)$

We continue with the abstract algebraic side of Drinfeld theory and also give some important examples. First of all, by applying S to Theorem 5.7, we have an immediate corollary,

Corollary 6.1 *The element* $\mathfrak{v} = S\mathfrak{u} \in H$ *obeys* $S^{-2}h = \mathfrak{v}h\mathfrak{v}^{-1}$ *for all* $h \in H$ *and* $\Delta\mathfrak{v} = Q^{-1}(\mathfrak{v} \otimes \mathfrak{v})$. *Hence* $\mathfrak{u}\mathfrak{v} = \mathfrak{v}\mathfrak{u}$ *is central and obeys* $\Delta(\mathfrak{u}\mathfrak{v}) = Q^{-2}(\mathfrak{u}\mathfrak{v} \otimes \mathfrak{u}\mathfrak{v})$. *The element* $\mathfrak{u}\mathfrak{v}^{-1} = \mathfrak{v}^{-1}\mathfrak{u}$ *is grouplike and implements* S^4 *by conjugation.*

Proof (i) We use the usual properties of the antipode. Clearly, Theorem 5.7 implies $S^2\mathfrak{u} = \mathfrak{u}$ and hence $S(\mathfrak{v}h\mathfrak{v}^{-1}) = S(S\mathfrak{u})^{-1}(Sh)S^2\mathfrak{u} = \mathfrak{u}^{-1}(Sh)\mathfrak{u} = S^{-1}h$ for all $h \in H$. Similarly $\Delta\mathfrak{v} = \Delta(S\mathfrak{u}) = \tau \circ (S \otimes S) \circ \Delta\mathfrak{u} = (\mathfrak{v} \otimes \mathfrak{v})\tau \circ (S \otimes S)(Q^{-1}) = (\mathfrak{v} \otimes \mathfrak{v})Q^{-1} = Q^{-1}(\mathfrak{v} \otimes \mathfrak{v})$, by S-invariance of $\mathcal{R}$.

(ii) Given this, $\mathfrak{u}\mathfrak{v} = \mathfrak{v}S^2(\mathfrak{u}) = \mathfrak{v}\mathfrak{u}$, etc. Also, $\mathfrak{u}\mathfrak{v}h = S^2(S^{-2}(h))\mathfrak{u}\mathfrak{v} = h\mathfrak{u}\mathfrak{v}$ for all h, so $\mathfrak{u}\mathfrak{v}$ is central. Moreover, $\Delta\mathfrak{u}\mathfrak{v} = Q^{-1}(\mathfrak{u} \otimes \mathfrak{u})Q^{-1}(\mathfrak{v} \otimes \mathfrak{v}) = Q^{-2}(\mathfrak{u}\mathfrak{v} \otimes \mathfrak{u}\mathfrak{v})$ by S-invariance of $\mathcal{R}$. Likewise, $\Delta\mathfrak{u}\mathfrak{v}^{-1} = \mathfrak{u}\mathfrak{v}^{-1} \otimes \mathfrak{u}\mathfrak{v}^{-1}$ and $\mathfrak{u}\mathfrak{v}^{-1}$ implements S^4 because $\mathfrak{u}, \mathfrak{v}^{-1}$ each implement S^2. $\qquad\square$

The automorphisms S^4 and S^2 play an important role in the theory of finite-dimensional Hopf algebras. For example, in characteristic zero $S^2 = \mathrm{id}$ *iff* the Hopf algebra is semisimple, in which case $\mathfrak{u}, \mathfrak{v}$ are separately central. These elements $\mathfrak{u}, \mathfrak{v}$ play an important role in the application of quantum groups to knot theory, as will see later.

Definition 6.2 *A quasitriangular Hopf algebra is called* ribbon *if* $\mathfrak{u}\mathfrak{v}$ *has a central square root* ν *(the ribbon element), such that*

$$\nu^2 = \mathfrak{v}\mathfrak{u}, \qquad \Delta\nu = Q^{-1}(\nu \otimes \nu), \qquad \epsilon\nu = 1, \qquad S\nu = \nu.$$

To give a concrete example demonstrating all of these ideas, let $\mathbb{Z}_{/n} \equiv \mathbb{Z}/n\mathbb{Z}$ be the finite cyclic group of order n. We consider $H = \mathbb{C}\mathbb{Z}_{/n}$, its group algebra over $\mathbb{C}$. This is $\mathbb{C}[g]$ modulo the relation $g^n = 1$, with

$$\Delta g = g \otimes g, \quad \epsilon g = 1, \quad Sg = g^{-1} = g^{n-1}.$$

Of course, it is trivially quasitriangular with $\mathcal{R} = 1 \otimes 1$. But we can also consider it with a different quasitriangular structure.

Example 6.3 *Let* $\mathbb{C}_q\mathbb{Z}_{/n}$ *denote the quasitriangular Hopf algebra consisting of* $\mathbb{C}\mathbb{Z}_{/n}$ *and*

$$\mathcal{R} = \frac{1}{n} \sum_{a,b=0}^{n-1} q^{-ab} g^a \otimes g^b$$

where $q = e^{\frac{2\pi i}{n}}$. *It is factorisable iff* n *is odd and* > 2 *(and triangular for* $n = 2$*), and ribbon. We have*

$$Q = \frac{1}{n} \sum_{a,b=0}^{n-1} q^{-ab} g^{2a} \otimes g^b, \quad \mathfrak{u} = \mathfrak{v} = \nu = \frac{1}{n} \sum_{a=0}^{n-1} g^a \theta(a)$$

where $\theta(a) = \sum_{b=0}^{n-1} q^{-(a+b)b}$ *is the* $\mathbb{Z}_{/n}$ *theta-function.*

Proof Note that $n^{-1} \sum_{b=0}^{n-1} q^{ab} = \delta_{a,0}$. Then

$$\begin{aligned}
\mathcal{R}_{13}\mathcal{R}_{23} &= \frac{1}{n^2} \sum_{a,b,c,d} q^{-(ab+cd)} g^a \otimes g^c \otimes g^{b+d} \\
&= \frac{1}{n^2} \sum_{a,b,c,d'} q^{-b(a-c)} q^{-cd'} g^a \otimes g^c \otimes g^{d'} \\
&= \frac{1}{n} \sum_{a,b} q^{-ab} g^a \otimes g^a \otimes g^b = (\Delta \otimes \mathrm{id})\mathcal{R},
\end{aligned}$$

where $d' = d + b$ is a change of variables. Similarly for the second half of axiom 3 of a quasitriangular structure. Axiom 2 is automatic because the Hopf algebra is both commutative and cocommutative. Next, we compute $Q = \mathcal{R}_{21}\mathcal{R} = \mathcal{R}^2$ here. This is

$$Q = \frac{1}{n^2} \sum_{a,b,c,d} q^{-ab-cd} g^{a+c} \otimes g^{b+d} = \frac{1}{n^2} \sum_{a',b',c,d} q^{-(a'-c)(b'-d)-cd} g^{a'} \otimes g^{b'}$$

where we change variables to $a' = a + c$, $b' = b + d$. We can now do the d summation, which yields $\delta_{a'-2c,0}$, and then the a' summation by setting $a' = 2c$. This gives the stated Q after a relabelling. Hence

$$\sum_c q^{cd} f^c(Q^{(1)}) Q^{(2)} = \frac{1}{n} \sum_{a,b,c} q^{-ab+cd} \delta_{c,2a} g^b = \frac{1}{n} \sum_{a,b} q^{-a(b-2d)} g^b = g^{2d}$$

for all d, where $\{f^c\}$ is a dual basis to the basis $\{g^a\}$ of $\mathbb{C}\mathbb{Z}_{/n}$. If n is odd, then $2d$ is a permutation of $\{0, \dots, n-1\}$ as d runs through this set, and otherwise not. Hence $\mathbb{Z}_{/n}$ is factorisable *iff* n is odd. The formulae for u, v are immediate from the formulae for u, v in the proof of the Theorem 5.7. One should also check that $\mathcal{R}$ is invertible. Indeed,

$$\mathcal{R}^{-1} = \frac{1}{n} \sum_{a,b=0}^{n-1} q^{ab} g^a \otimes g^b$$

as one may easily check by a computation similar to that for Q. □

This example generates as its category of modules a very interesting braided category, as we will see later in the course. We are now ready to introduce the famous Hopf algebra $u_q(sl_2)$.

Example 6.4 *Let q be a primitive n'th root of unity and $n > 2$ odd. The quasitriangular Hopf algebra $H = u_q(sl_2)$ over $\mathbb{C}$ is $\mathbb{C}\langle E, F, g, g^{-1} \rangle$ modulo the relations*

$$gEg^{-1} = q^2 E, \quad gFg^{-1} = q^{-2} F, \quad [E, F] = \frac{g - g^{-1}}{q - q^{-1}},$$

$$g^n = 1, \quad E^n = F^n = 0,$$

with the coalgebra and antipode

$$\Delta g = g \otimes g, \quad \Delta E = E \otimes g + 1 \otimes E, \quad \Delta F = F \otimes 1 + g^{-1} \otimes F$$

$$\epsilon(g) = 1, \quad \epsilon(E) = \epsilon(F) = 0, \quad Sg = g^{-1}, \quad SE = -Eg^{-1}, \quad SF = -gF$$

and the quasitriangular structure

$$\mathcal{R} = \mathcal{R}_g e_{q^{-2}}^{(q - q^{-1}) E \otimes F},$$

where

$$\mathcal{R}_g = \frac{1}{n} \sum_{a,b=0}^{n-1} q^{-2ab} g^a \otimes g^b, \quad e_{q^{-2}}^{(q-q^{-1}) E \otimes F} = \sum_{r=0}^{n-1} \frac{(q - q^{-1})^r}{[r]_{q^{-2}}!} E^r \otimes F^r,$$

and $[r]_{q^{-2}} = \frac{1 - q^{-2r}}{1 - q^{-2}}$.

Proof We check first that we have a Hopf algebra when Δ, ϵ, S are defined as stated on the generators and extended (anti)multiplicatively to products. The relation $gEg^{-1} = q^2E$ and the coproduct on E, g are similar to the version of $U_q(b_+)$ which we have already checked in Lecture 1. Similarly for $gFg^{-1} = q^{-2}F$ and the coproduct on F, g. We check next the $[E, F] = (g - g^{-1})/(q - q^{-1})$ relation,

$$\Delta[E, F] = (E \otimes g + 1 \otimes E)(F \otimes 1 + g^{-1} \otimes F)$$
$$-(F \otimes 1 + g^{-1} \otimes F)(E \otimes g + 1 \otimes E)$$
$$= \frac{(g - g^{-1})}{q - q^{-1}} \otimes g + Eg^{-1} \otimes gF - g^{-1}E \otimes Fg + g^{-1} \otimes \frac{(g - g^{-1})}{q - q^{-1}}$$
$$= \frac{1}{q - q^{-1}}(g \otimes g - g^{-1} \otimes g^{-1}) = \Delta \frac{(g - g^{-1})}{q - q^{-1}}.$$

So Δ descends to the coproduct defined by all these three relations. The check for ϵ and S is easy as well. We call the Hopf algebra with these three relations $U_q(sl_2)$ (capital U); we have not yet used that q is a root of unity, so this all works over a general field k with $q \in k^*$ such that $q^2 \neq 1$.

Next we further quotient this Hopf algebra by the relation $E^n = 0$. We need the formula, similar to the one we quoted already for the version of $U_q(b_+)$ in Lecture 1,

$$\Delta E^m = \sum_{r=0}^{m} \begin{bmatrix} m \\ r \end{bmatrix}_{q^{-2}} E^r \otimes g^r E^{m-r}.$$

This follows from the q-binomial formula proven in Lemma 7.1 in the next lecture, in the present case with $A = E \otimes g$, $B = 1 \otimes E$ so that $AB = q^2BA$ by the relations $gEg^{-1} = q^2E$. In particular

$$\Delta E^n = E^n \otimes g^n + 1 \otimes E^n$$

since only $r = 0, n$ contribute in this case; the other q-binomial coefficients have a factor $[n]_{q^{-2}}$, which vanishes as $q^n = 1$. None of the lower q-integers vanish since q is odd and $q^m \neq 1$ for $m < n$ (this is what we mean by a primitive n'th root). Therefore, we can set $E^n = 0$. Similarly for $F^n = 0$. The relation $g^n = 1$ is clearly compatible with Δ. This completes the proof that $u_q(sl_2)$ is a Hopf algebra.

Finally, we turn to the quasitriangular structure. For the quasico-commutativity axiom 2 on ΔE, we need $\mathcal{R}(E \otimes g + 1 \otimes E) = (E \otimes 1 + g \otimes E)\mathcal{R}$. But $\mathcal{R}_g(1 \otimes E) = (g \otimes E)\mathcal{R}_g$ by the $gE = q^2Eg$ relations and a change of variable in the sum for $\mathcal{R}_g$. Similarly for $E \otimes 1$ since $\mathcal{R}_g$ is symmetric. Hence

$$(E \otimes 1 + g \otimes E)\mathcal{R}_g = \mathcal{R}_g(E \otimes g^{-1} + 1 \otimes E).$$

In view of this, we require

$$e_{q^{-2}}^A(B + C) = (C + D)e_{q^{-2}}^A,$$

where $A = (q - q^{-1})E \otimes F$, $B = E \otimes g$, $C = 1 \otimes E$, $D = E \otimes g^{-1}$. This is proven in Lemma 7.2 in the next lecture for elements of any algebra obeying $AB = q^2BA$, $AD = q^{-2}DA$, $[C, A] = B - D$. In our case, these hold in the tensor product algebra via the algebra relations. Similarly for ΔF. For Δg we note that $g \otimes g$ commutes with A and hence with $e_{q^{-2}}^A$, where $A = (q - q^{-1})E \otimes F$. For the quasitriangular structure axiom 3, we note that this already holds for $\mathcal{R}_g$ since this is the quasitriangular structure in $\mathbb{C}_{q^2}\mathbb{Z}_{/n}$, which appears as a sub-Hopf algebra and which we have already verified above. We also have

$$(\Delta \otimes \mathrm{id})e_{q^{-2}}^{(q-q^{-1})E \otimes F} = e_{q^{-2}}^{(q-q^{-1})(E \otimes g \otimes F + 1 \otimes E \otimes F)}$$
$$= e_{q^{-2}}^{(q-q^{-1})E \otimes g \otimes F} e_{q^{-2}}^{(q-q^{-1})1 \otimes E \otimes F}$$

by the q-binomial lemma again, with $A = E \otimes g \otimes F$ and $B = 1 \otimes E \otimes F$ obeying $AB = q^2BA$ in the triple tensor product (then $e_{q^{-2}}^{A+B} = e_{q^{-2}}^A e_{q^{-2}}^B$ by more or less the usual argument). Then

$$(\Delta \otimes \mathrm{id})\mathcal{R} = \mathcal{R}_{g13}\mathcal{R}_{g23}e_{q^{-2}}^{(q-q^{-1})E \otimes g \otimes F} e_{q^{-2}}^{(q-q^{-1})1 \otimes E \otimes F} = \mathcal{R}_{13}\mathcal{R}_{23},$$

where we use the $gFg^{-1} = q^{-2}F$ relation to move $\mathcal{R}_{g23}$ to the right, which also has the effect of turning $E \otimes g \otimes F$ into $E \otimes 1 \otimes F$ as required. Here $\mathcal{R}_g(g \otimes F) = (1 \otimes F)\mathcal{R}_g$ along the same lines as the similar identity for $1 \otimes E$ already used.

We need one more detail to complete the proof, namely that $\mathcal{R}$ is invertible. Actually, the inverse is

$$\mathcal{R}^{-1} = e_{q^2}^{-(q-q^{-1})E \otimes F}\mathcal{R}_g^{-1}$$

by an elementary property of q-exponentials which we will see later. However, for the present it is enough to check that $(\epsilon \otimes \mathrm{id})\mathcal{R} = 1$, which is immediate in our case; as we have seen in the proof of Lemma 5.3, $\mathcal{R}(S \otimes \mathrm{id})\mathcal{R} = (\epsilon \otimes \mathrm{id})(\mathcal{R}) = ((S \otimes \mathrm{id})\mathcal{R})\mathcal{R}$ once axiom 3 of a quasitriangular structure holds, so we can take $\mathcal{R}^{-1} = (S \otimes \mathrm{id})\mathcal{R}$ as a definition. $\square$

The reader is still owed some standard lemmas from q-analysis, which we will do in the next lecture.

7

q-Binomials

We prove some lemmas from what is called 'q-analysis', which were quoted during the construction of the quantum group $u_q(sl_2)$ in the last lecture. To apply them one should change q to q^{-2}. However, the lemmas are used elsewhere as well, so we present them in their natural form.

Lemma 7.1 *(q-Binomial formula) Let $q \in k^*$. Let*

$$\begin{bmatrix} n \\ m \end{bmatrix}_q = \frac{[n]_q!}{[m]_q![n-m]_q!}, \quad [m]_q = \frac{1-q^m}{1-q}$$

where we suppose (for convenience) that the q-integers $[m]_q$ are nonzero for $0 < m < n$. By convention, they are defined as the usual integers when $q = 1$. Also by convention, $\begin{bmatrix} n \\ m \end{bmatrix}_q = 1$ when $m = 0, n$. If A, B are elements of an algebra obeying $BA = qAB$ then

$$(A+B)^n = \sum_{m=0}^{n} \begin{bmatrix} n \\ m \end{bmatrix}_q A^m B^{n-m}.$$

Proof We proceed by induction. Assuming the result for $(A+B)^{n-1}$, we have

$$(A+B)^{n-1}(A+B) = \sum_{m=0}^{n-1} \begin{bmatrix} n-1 \\ m \end{bmatrix}_q A^m B^{n-1-m}(A+B)$$

$$= \sum_{m=0}^{n-1} q^{n-1-m} \begin{bmatrix} n-1 \\ m \end{bmatrix}_q A^{m+1} B^{n-1-m} + \sum_{m=0}^{n-1} \begin{bmatrix} n-1 \\ m \end{bmatrix}_q A^m B^{n-m}$$

$$= \sum_{m=1}^{n} q^{n-m} \begin{bmatrix} n-1 \\ m-1 \end{bmatrix}_q A^m B^{n-m} + \sum_{m=0}^{n-1} \begin{bmatrix} n-1 \\ m \end{bmatrix}_q A^m B^{n-m}$$

$$= A^n + B^n + \sum_{m=1}^{n-1} \left(q^{n-m} \begin{bmatrix} n-1 \\ m-1 \end{bmatrix}_q + \begin{bmatrix} n-1 \\ m \end{bmatrix}_q \right) A^m B^{n-m}.$$

The expression in parentheses combines to $\begin{bmatrix} n \\ m \end{bmatrix}_q$, as required, after an elementary computation using the identity

$$q^{n-m}[m]_q + [n-m]_q = [n]_q$$

(the addition rule for q-integers).   □

We used this in the expression for ΔE^m in $u_q(sl_2)$. Similarly, it implies the formula for the coproduct of the version of $U_q(b_+)$ in Lecture 1. Also, one does not actually need the q-integers here to be invertible. The easiest way to see this is to define the q-binomial coefficients inductively by

$$\begin{bmatrix} n \\ m \end{bmatrix}_q = q^{n-m} \begin{bmatrix} n-1 \\ m-1 \end{bmatrix}_q + \begin{bmatrix} n-1 \\ m \end{bmatrix}_q, \quad \begin{bmatrix} n \\ 0 \end{bmatrix}_q = 1,$$

and zero when $m > n$.

As an immediate corollary, let

$$e_q^A = \sum_{m=0}^{\infty} \frac{A^m}{[m]_q!}$$

be the q-exponential. One can treat this as a formal powerseries, but in our case we suppose that A is nilpotent, so that the series is actually finite, and that the $[m]_q$ are invertible for m less than the nilpotency degree N. Given another element B obeying $BA = qAB$ and jointly nilpotent with A in the sense $A^m B^n = 0$ whenever $n + m = N$, we have

$$e_q^{A+B} = e_q^A e_q^B.$$

The computation is exactly the same as the usual one but with the q-binomial formula in place of the usual binomial formula, and higher terms in the usual powerseries for the exponential function dropping out by the nilpotency. This is what we used in proving the quasitriangular structure for $u_q(sl_2)$, where the nilpotency holds because both A, B as stated there have the factor $\otimes F$. This is the additional subtlety beyond the relations required to apply the lemma.

Lemma 7.2 Let $q \in k^*$ and suppose that A, B, C, D are elements of an algebra obeying $BA = qAB$, $AD = qDA$ and $[C, A] = B - D$. Then

$$[C, A^n] = [n]_q(A^{n-1}B - DA^{n-1})$$

Proof We again proceed by induction. Assuming the result for A^{n-1}, we have

$$
\begin{aligned}
[C, A^n] &= A[C, A^{n-1}] + (B - D)A^{n-1} \\
&= A[n-1]_q(A^{n-2}B - DA^{n-2}) + (B - D)A^{n-1} \\
&= ([n-1]_q + q^{n-1})A^{n-1}B - (1 + q[n-1]_q)DA^{n-1} \\
&= [n]_q(A^{n-1}B - DA^{n-1}),
\end{aligned}
$$

where we have used $[n-1]_q + q^{n-1} = [n]_q = 1 + q[n-1]_q$ as a special case of the addition rule for q-integers. $\qquad\square$

Applying the lemma to the q-exponentials immediately gives

$$
[C, e_q^A] = e_q^A B - De_q^A,
$$

which, with a suitable choice of q, is the identity used in the last lecture. Again, the series are finite since A was nilpotent.

Proposition 7.3 *The quantum group $u_q(sl_2)$ is factorisable.*

Proof We write $\mathcal{R}_g e_{q^{-2}}^{(q-q^{-1})E \otimes F} = e_{q^{-2}}^{(q-q^{-1})Eg^{-1} \otimes gF} \mathcal{R}_g$ by the identities in the proof of Example 6.4. Then $Q = e_{q^{-2}}^{(q-q^{-1})gF \otimes Eg^{-1}} Q_g e_{q^{-2}}^{(q-q^{-1})E \otimes F}$ $= \sum_{a,b} \alpha_a \alpha_b (gF)^a Q_g^{(1)} E^b \otimes (Eg^{-1})^a Q_g^{(2)} F^b$, where $Q_g = \mathcal{R}_g^2$ and the α_a are the coefficients from $e_{q^{-2}}$. A basis of $u_q(sl_2)$ is provided either by expressions of the form $\{(Eg^{-1})^a g^c F^b\}$ or by expressions of the form $\{(gF)^a g^c E^b\}$. Hence, by considering elements of the dual basis to the latter, evaluation against $Q^{(1)}$ is surjective *iff* evaluation with $Q_g^{(1)}$ is surjective, i.e. *iff* $\mathbb{C}_{q^2}\mathbb{Z}/n$ is factorisable. This is the case when n is odd and $n > 1$ as it is here, by a slight refinement of the proof for $\mathbb{C}_q\mathbb{Z}/n$. $\qquad\square$

Actually, there are at least two nonisomorphic versions of $u_q(sl_2)$. For example, the historically first one has strange relations like $g^{4n} = 1$ and is *not* factorisable. The quantum group $u_q(sl_2)$ can also be shown to be ribbon. An explicit formula for the ribbon element is

$$
\nu = \frac{1}{n} \left(\sum_{m=0}^{n-1} q^{2m^2} \right) \sum_{m,a=0}^{n-1} \frac{(q-q^{-1})^m}{[m]_{q^{-2}}!} q^{-\frac{1}{2}(n+1)(m-a-1)^2} E^m g^a F^m.
$$

Before proceeding to braided categories and knots, let us summarise some aspects of the construction of $u_q(sl_2)$. First of all, in the proof of its Hopf algebra structure we first constructed the Hopf algebra $U_q(sl_2)$

which works over a general field with $q \in k^*$ and $q^2 \neq 1$. This is $k\langle E, F, g, g^{-1} \rangle$ modulo the relations

$$gEg^{-1} = q^2 E, \quad gFg^{-1} = q^{-2}F, \quad [E, F] = \frac{g - g^{-1}}{q - q^{-1}}.$$

The coproduct is

$$\Delta E = E \otimes g + 1 \otimes E, \quad \Delta F = F \otimes 1 + g^{-1} \otimes F, \quad \Delta g = g \otimes g.$$

Clearly the g, g^{-1}, E form a sub-Hopf algebra, which can be denoted $U_q(b_+)$ (it is a variant of the one in Lecture 1). Similarly, g, g^{-1}, F form a sub-Hopf algebra which can be denoted $U_q(b_-)$. It is high time to explain these notations, which we do now.

In fact, there are several variants even of this Hopf algebra $U_q(sl_2)$, not only by changing the exact choice of generators and relations, but more fundamentally by changing how we view q. In one variant (Lusztig) one regards q as an indeterminate and considers these relations as defining a Hopf algebra over the field $\mathbb{Q}(q)$ of rational functions in q. One may go further and define a version of it over the commutative ring $\mathbb{Q}[q, q^{-1}]$ or even over $\mathbb{Z}[q, q^{-1}]$, with care. Until now we have studied all Hopf algebras over a field, but the diagrammatic definitions make sense in much greater generality, in particular over a commutative ring.

Finally, (Drinfeld's original approach) one can write $q = e^{\frac{t}{2}}$, and work over the ring of formal powerseries $\mathbb{C}[[t]]$. Here t is called a deformation parameter and by expanding expressions in powers of t we can effectively see how the structure behaves near $q = 1$ (i.e. $t = 0$). [So t plays a role similar to Planck's constant $\hbar$ in physics.] In this variant the generators are H, X_+, X_-, with the relations and structure

$$[H, X_\pm] = \pm 2 X_\pm, \quad [X_+, X_-] = \frac{q^H - q^{-H}}{q - q^{-1}}$$

$$\Delta H = H \otimes 1 + 1 \otimes H, \quad \Delta X_\pm = X_\pm \otimes q^{\frac{H}{2}} + q^{-\frac{H}{2}} \otimes X_\pm$$

$$\mathcal{R} = q^{\frac{H \otimes H}{2}} \sum_{n=0}^{\infty} \frac{(1 - q^{-2})^n}{q[n]!} \left(q^{\frac{H}{2}} X_+ \otimes q^{-\frac{H}{2}} X_- \right)^n q^{\frac{n(n-1)}{2}},$$

$$q[n] \equiv \frac{q^n - q^{-n}}{q - q^{-1}}.$$

Here $q = e^{\frac{t}{2}}$ should be substituted in all expressions. We also use here some other popular conventions where the q-integers are more symmetrical between q and q^{-1}. One can check that if

$$g = q^H, \quad E = X_+ q^{\frac{H}{2}}, \quad F = q^{-\frac{H}{2}} X_-$$

then these obey the relations of $U_q(sl_2)$ given before. Moreover, and most importantly, we have the quasitriangular structure once again, as a formal powerseries in t.

This version of $U_q(sl_2)$ (also denoted $U_t(sl_2)$) is a deformation in the sense that as a $\mathbb{C}[\![t]\!]$ module it coincides with $U(sl_2)[\![t]\!]$ (formal powerseries in t with coefficients in the classical enveloping algebra $U(sl_2)$). In particular, if we look at the algebra relations to the lowest order in t, they are

$$[H, X_\pm] = \pm 2 X_\pm, \quad [X_+, X_-] = H,$$

which are the relations of the enveloping algebra of the Lie algebra sl_2. The coproduct

$$\Delta X_\pm = X_\pm \otimes 1 + 1 \otimes X_\pm + \frac{t}{4}(X_\pm \otimes H - H \otimes X_\pm) + \cdots$$

also reduces at the lowest order to the usual one for $U(sl_2)$. The next order term in the coproduct defines a map

$$\delta : sl_2 \to sl_2 \otimes sl_2$$

obeying axioms which are an infinitesimal version of a bialgebra (called a Lie bialgebra). We will come back to this as a special topic near the end of the course. In the meantime, the reader can have some fun reversing all the arrows of a Lie algebra (a Lie coalgebra) and thinking about what the homomorphism property of Δ becomes at this Lie algebra level.

8

Quantum double. Dual quasitriangular structures

In this lecture we finish up our basic Hopf algebra theory with some general constructions. First of all, we pulled the structure of $u_q(sl_2)$ somewhat 'out of a hat'. Although it is standard enough that one should work with its structure directly, it is in fact an example of a general construction called the 'quantum double'.

Proposition 8.1 *(Quantum double) Let H be a finite-dimensional Hopf algebra. Then the vector space $H^* \otimes H$ has the following structure of a Hopf algebra $D(H)$ (the quantum double of H, also denoted $H^{*op} \bowtie H$). Here*

$$(\phi \otimes h)(\psi \otimes g) = \sum \psi_{(2)} \phi \otimes h_{(2)} g \langle Sh_{(1)}, \psi_{(1)} \rangle \langle h_{(3)}, \psi_{(3)} \rangle$$

for all $\phi, \psi \in H^*$ and $h, g \in H$, with unit $1 \otimes 1$ and the coalgebra

$$\Delta(\phi \otimes h) = \sum (\phi_{(1)} \otimes h_{(1)}) \otimes (\phi_{(2)} \otimes h_{(2)}), \quad \epsilon(\phi \otimes h) = \epsilon(\phi)\epsilon(h)$$

for all $\phi \in H^*$ and $h \in H$. Here H, H^{*op} (H^* with the opposite product) appear as sub-Hopf algebras.

Proof It is an exercise, as in Problem I.21, to verify that the stated product is indeed associative. Assuming this, we verify that the tensor product coproduct is an algebra homomorphism:

$$\Delta(\phi \otimes h)\Delta(\psi \otimes g) = (\phi_{(1)} \otimes h_{(1)})(\psi_{(1)} \otimes g_{(1)}) \otimes (\phi_{(2)} \otimes h_{(2)})(\psi_{(2)} \otimes g_{(2)})$$

$$= \psi_{(2)} \phi_{(1)} \otimes h_{(2)} g_{(1)} \otimes \psi_{(5)} \phi_{(2)} \otimes h_{(5)} g_{(2)} \langle Sh_{(1)}, \psi_{(1)} \rangle \langle h_{(3)}, \psi_{(3)} \rangle$$
$$\langle Sh_{(4)}, \psi_{(4)} \rangle \langle h_{(6)}, \psi_{(6)} \rangle$$

$$= \psi_{(2)} \phi_{(1)} \otimes h_{(2)} g_{(1)} \otimes \psi_{(3)} \phi_{(2)} \otimes h_{(3)} g_{(2)} \langle Sh_{(1)}, \psi_{(1)} \rangle \langle h_{(4)}, \psi_{(4)} \rangle$$

$$= \Delta((\phi \otimes h)(\psi \otimes g)).$$

44

The unit and counit are easily checked, so that we have a bialgebra. Next, we define

$$S(\phi \otimes h) = (1 \otimes Sh)(S^{-1}\phi \otimes 1) = S^{-1}\phi_{(2)} \otimes Sh_{(2)} \langle h_{(1)}, \phi_{(1)} \rangle \langle Sh_{(3)}, \phi_{(3)} \rangle,$$

and check that

$$(S(\phi_{(1)} \otimes h_{(1)}))(\phi_{(2)} \otimes h_{(2)}) = (1 \otimes Sh_{(1)})(S^{-1}\phi_{(1)} \otimes 1)(\phi_{(2)} \otimes 1)(1 \otimes h_{(2)})$$
$$= \epsilon(h)\epsilon(\phi),$$
$$(\phi_{(1)} \otimes h_{(1)})S(\phi_{(2)} \otimes h_{(2)}) = (\phi_{(1)} \otimes h_{(1)})(1 \otimes Sh_{(2)})(S^{-1}\phi_{(2)} \otimes 1)$$
$$= \epsilon(h)\epsilon(\phi).$$

The restrictions to H and $H^{*\mathrm{op}}$ here are S (so H appears as a sub-Hopf algebra) and S^{-1}, the inverse of the antipode of H^*. It is easy to see that a Hopf algebra taken with the opposite product remains a bialgebra but an antipode for the opposite product is the same thing as an inverse for the antipode of the original Hopf algebra. One can make a similar observation with regard to the opposite coproduct as the underlying reason for Theorem 5.7 on S^{-1}. In the present case, since H is finite dimensional, one can show that S^{-1} always exists. Otherwise (e.g. in the infinite-dimensional case) this should be imposed as an additional condition. $\qquad\qquad\Box$

The importance of this construction is that it ensures a plentiful supply of factorisable quasitriangular Hopf algebras:

Proposition 8.2 *If H is finite dimensional then $D(H)$ is a factorisable quasitriangular Hopf algebra, with*

$$\mathcal{R} = \sum_a (f^a \otimes 1) \otimes (1 \otimes e_a),$$

where $\{e_a\}$ is a basis of H and $\{f^a\}$ is a dual basis.

Proof We adopt the convention where repeated indices are summed. Then, to verify axiom 3 of a quasitriangular structure, we need

$$(f^a_{\ (1)} \otimes 1) \otimes (f^a_{\ (2)} \otimes 1) \otimes (1 \otimes e_a) = (f^a \otimes 1) \otimes (f^b \otimes 1) \otimes (1 \otimes e_a e_b),$$

$$(f^a \otimes 1) \otimes (1 \otimes e_{a(1)}) \otimes (1 \otimes e_{a(2)}) = (f^b f^a \otimes 1) \otimes (1 \otimes e_b) \otimes (1 \otimes e_a),$$

which are easily verified by evaluating against general elements: evaluating against $\phi \in H^*$ in the third factor gives both sides of the first identity as $\phi_{(1)} \otimes 1 \otimes \phi_{(2)} \otimes 1 \otimes 1$. Here $f^a \langle e_a, \phi \rangle = \phi$ and $f^a \otimes f^b \langle e_a e_b, \phi \rangle =$

$\phi_{(1)} \otimes \phi_{(2)}$. Similarly on evaluating against $h \in H$ for the second identity. For the quasicocommutativity axiom of a quasitriangular structure, we have

$$\begin{aligned}
\mathcal{R}\Delta(\phi \otimes h) &= (f^a \otimes 1)(\phi_{(1)} \otimes h_{(1)}) \otimes (1 \otimes e_a)(\phi_{(2)} \otimes h_{(2)}) \\
&= \phi_{(1)} f^a \otimes h_{(1)} \otimes \phi_{(3)} \otimes e_{a(2)} h_{(2)} \langle Se_{a(1)}, \phi_{(2)} \rangle \langle e_{a(3)}, \phi_{(4)} \rangle, \\
(\tau \circ \Delta(\phi \otimes h))\mathcal{R} &= (\phi_{(2)} \otimes h_{(2)})(f^a \otimes 1) \otimes (\phi_{(1)} \otimes h_{(1)})(1 \otimes e_a) \\
&= f^a{}_{(2)} \phi_{(2)} \otimes h_{(3)} \otimes \phi_{(1)} \otimes h_{(1)} e_a \langle Sh_{(2)}, f^a{}_{(1)} \rangle \langle h_{(4)}, f^a{}_{(3)} \rangle.
\end{aligned}$$

Evaluating against a general element $g \in H$ in the first tensor factor gives the first expression as

$$\begin{aligned}
\langle g, \mathcal{R}\Delta(\phi \otimes h) \rangle &= \langle g_{(1)}, \phi_{(1)} \rangle h_{(1)} \otimes \phi_{(3)} \otimes g_{(3)} h_{(2)} \langle Sg_{(2)}, \phi_{(2)} \rangle \langle g_{(4)}, \phi_{(4)} \rangle \\
&= h_{(1)} \otimes \phi_{(1)} \otimes g_{(1)} h_{(2)} \langle g_{(2)}, \phi_{(2)} \rangle, \\
&= \langle g_{(2)}, \phi_{(2)} \rangle h_{(3)} \otimes \phi_{(1)} \otimes h_{(1)} e_a \langle (Sh_{(2)}) g_{(1)} h_{(4)}, f^a \rangle \\
&= \langle g, (\tau \circ \Delta(\phi \otimes h))\mathcal{R} \rangle
\end{aligned}$$

as required. Finally, we need $\mathcal{R}$ to be invertible. In view of Lemma 5.3, we define

$$\mathcal{R}^{-1} = S^{-1} f^a \otimes 1 \otimes 1 \otimes e_a.$$

That this is an inverse follows from the other quasitriangularity axioms and $(\epsilon \otimes \mathrm{id})\mathcal{R} = 1$, which is clear.

For the factorisability, we compute the induced map $Q : H \otimes H^* \to H^* \otimes H$ as

$$\begin{aligned}
Q(h \otimes \phi) &= \langle h \otimes \phi, (1 \otimes e_a)(f^b \otimes 1) \rangle (f^a \otimes 1)(1 \otimes e_b) \\
&= \langle h, f^b{}_{(2)} \rangle \langle e_{a(2)}, \phi \rangle \langle Se_{a(1)}, f^b{}_{(1)} \rangle \langle e_{a(3)}, f^b{}_{(3)} \rangle f^a \otimes e_b \\
&= f^a \otimes (Se_{a(1)}) h e_{a(3)} \langle e_{a(2)}, \phi \rangle = f^a \phi f^b \otimes (Se_a) h e_b.
\end{aligned}$$

This is easily seen to be surjective as required. Indeed, there is an inverse map provided by

$$Q^{-1}(\phi \otimes h) = e_{a(1)} h Se_{a(3)} \otimes f^a \langle \phi, e_{a(2)} \rangle = e_a h Se_b \otimes f^a \phi f^b.$$

$\square$

For example, if we take $u_q(b_+)$ the sub-Hopf algebra generated by g, E in $u_q(sl_2)$ then we can identify $u_q(b_+)^{*\mathrm{op}} = u_q(b_-)$, the sub-Hopf algebra generated by g, F. This is a version of the self-duality pairing of $U_q(b_+)$ with itself which we have discussed in Lecture 2. Denoting the generators of $u_q(b_+)^{*\mathrm{op}}$ by g', F say, we can recover $u_q(sl_2)$ as $D(u_q(b_+))$

modulo (the ideal generated by) the relation $g = g'$. Its quasitriangular structure is then the quotient of that of $D(u_q(b_+))$.

Finally, to complete our algebraic picture, we should reverse all arrows and arrive at the dual version of Drinfeld theory, i.e. a theory of dual quasitriangular or 'coquasitriangular' Hopf algebras. We do not need to redo all proofs as long as we dualise correctly. The idea this time is to describe Hopf algebras H which are noncommutative but for which the noncommutativity is under control by a map $H \otimes H \to k$. We think of a quasitriangular structure as a map $k \to H \otimes H$ and reverse the arrows. The first step is the to find the correct notion of $\mathcal{R}^{-1}$.

Lemma 8.3 *Let C be a coalgebra and A an algebra. Then the set of linear maps* $\operatorname{Hom}(C, A)$ *has an associative product and unit*

$$\phi\psi = m \circ (\phi \otimes \psi) \circ \Delta, \quad I = \eta \circ \epsilon.$$

This is called the 'convolution algebra'. Explicitly,

$$(\phi\psi)(c) = \sum \phi(c_{(1)})\psi(c_{(2)}), \quad I(c) = 1\epsilon(c), \quad \forall c \in C.$$

Proof Elementary. □

It is in this algebra that we require $\mathcal{R}^{-1} : H \otimes H \to k$ to exist. Explicitly, it means a linear map such that

$$\sum \mathcal{R}^{-1}(h_{(1)} \otimes g_{(1)})\mathcal{R}(h_{(2)} \otimes g_{(2)})$$
$$= \epsilon(h)\epsilon(g) = \sum \mathcal{R}(h_{(1)} \otimes g_{(1)})\mathcal{R}^{-1}(h_{(2)} \otimes g_{(2)})$$

for all $h, g \in H$. Keeping such considerations in mind, it is easy to dualise the remainder of Drinfeld's axioms to obtain the following definition.

Definition 8.4 *A dual quasitriangular (or coquasitriangular) bialgebra or Hopf algebra is a pair $(H, \mathcal{R})$ where*

1. *H is a bialgebra or Hopf algebra.*
2. *$\mathcal{R}$ is a convolution-invertible map $\mathcal{R} : H \otimes H \to k$ such that*

$$\sum g_{(1)}h_{(1)}\mathcal{R}(h_{(2)} \otimes g_{(2)}) = \sum \mathcal{R}(h_{(1)} \otimes g_{(1)})h_{(2)}g_{(2)}$$

for all $h, g \in H$ (quasicommutativity axiom).

3.

$$\mathcal{R}(hg \otimes f) = \sum \mathcal{R}(h \otimes f_{(1)})\mathcal{R}(g \otimes f_{(2)}),$$
$$\mathcal{R}(h \otimes gf) = \sum \mathcal{R}(h_{(1)} \otimes f)\mathcal{R}(h_{(2)} \otimes g)$$

for all $h, g, f \in H$ (bicharacter axioms).

We state without full proof the arrow-reversed versions of Lemma 5.2, Lemma 5.3 and Theorem 5.7. It is an exercise to write out the explicit proof more fully if one wants to.

Lemma 8.5 *If $(H, \mathcal{R})$ is a dual quasitriangular bialgebra, then*
1. *$\mathcal{R}(h \otimes 1) = \epsilon(h) = \mathcal{R}(1 \otimes h), \quad \forall h \in H$.*
2. *The Yang–Baxter equation holds in the form*

$$\sum \mathcal{R}(h_{(1)} \otimes g_{(1)})\mathcal{R}(h_{(2)} \otimes f_{(1)})\mathcal{R}(g_{(2)} \otimes f_{(2)})$$
$$= \sum \mathcal{R}(g_{(1)} \otimes f_{(1)})\mathcal{R}(h_{(1)} \otimes f_{(2)})\mathcal{R}(h_{(2)} \otimes g_{(2)})$$

for all $h, g, f \in H$.
3. *In the Hopf algebra case*

$$\mathcal{R}(Sh \otimes g) = \mathcal{R}^{-1}(h \otimes g), \qquad \mathcal{R}^{-1}(h \otimes Sg) = \mathcal{R}(h \otimes g),$$

$$\mathcal{R}(Sh \otimes Sg) = \mathcal{R}(h \otimes g), \quad \forall h, g \in H.$$

Proof For part 2, expand $\mathcal{R}(h \otimes f_{(1)}g_{(1)})\mathcal{R}(g_{(2)} \otimes f_{(2)})$ in two ways, either using the quasicommutativity axiom first and then the bicharacter axiom, or the bicharacter axiom directly. $\qquad \square$

Similarly, one has

Theorem 8.6 *Let $(H, \mathcal{R})$ be a dual quasitriangular Hopf algebra. Then S is invertible and there is a convolution-invertible map $\mathfrak{v} : H \to k$ such that*

$$\sum h_{(1)}\mathfrak{v}(h_{(2)}) = \sum \mathfrak{v}(h_{(1)})S^2 h_{(2)}, \quad \forall h \in H.$$

Proof We do not want to write out all the proofs again (just reverse all arrows); suffice it to say that

$$\mathfrak{v}(h) = \mathcal{R}(h_{(1)} \otimes Sh_{(2)}), \qquad \mathfrak{v}^{-1}(h) = \mathcal{R}(S^2 h_{(1)} \otimes h_{(2)})$$

as dual to the constructions in the theorem for quasitriangular Hopf algebras. Actually, the dual is the corresponding theorem for

$$\mathfrak{u}(h) = \mathcal{R}(h_{(2)} \otimes Sh_{(1)}), \qquad \mathfrak{u}^{-1}(h) = \mathcal{R}(S^2 h_{(2)} \otimes h_{(1)})$$

obeying $\mathfrak{u}(h_{(1)})h_{(2)} = S^2 h_{(1)}\mathfrak{u}(h_{(2)})$ so we are also making a left–right reversal. $\qquad \square$

Other aspects of Drinfeld theory can clearly be dualised in just the same way. Thus $\mathfrak{u}, \mathfrak{v}$ are now almost multiplicative, up to the linear functional on $H \otimes H$ given by

$$Q(h \otimes g) = \sum \mathcal{R}(g_{(1)} \otimes h_{(1)}) \mathcal{R}(h_{(2)} \otimes g_{(2)}), \quad \forall h, g \in H.$$

A dual quasitriangular Hopf algebra is dual triangular if $Q(h \otimes g) = \epsilon(h)\epsilon(g)$. It is 'factorisable' if Q is nondegenerate in the sense that $Q(h \otimes g) = 0$ for all g implies that $h = 0$. And a dual quasitriangular Hopf algebra is ribbon if the linear functional $\mathfrak{u}\mathfrak{v}(h) = \sum \mathfrak{u}(h_{(1)})\mathfrak{v}(h_{(2)})$ has a square root etc., in the convolution algebra of maps $H \to k$.

As with comodules, working with these dual quasitriangular structures often keeps things algebraic. For example, the group algebra is defined for any group G, not only finite ones. Clearly in this case the definition of a dual quasitriangular structure reduces to $\mathcal{R}$ being the extension by linearity of a bicharacter $\mathcal{R} : G \times G \to k$ and the requirement that G be Abelian.

Example 8.7 *Let $q \in k^*$. We denote by $k_q\mathbb{Z}$ the group algebra $k\mathbb{Z} = k[g, g^{-1}]$ of $\mathbb{Z}$ equipped with the dual quasitriangular structure defined by bicharacter*

$$\mathcal{R}(g^m \otimes g^n) = q^{mn}.$$

Proof The algebra is commutative and $\Delta g = g \otimes g$, so the quasicommutativity axiom is automatic. The bicharacter axiom is immediate. Convolution-invertibility reduces to invertibility of q. Since $\mathbb{Z}$ is free, this is in fact the only possibility for a dual quasitriangular structure on $k\mathbb{Z}$. $\qquad\square$

Similarly, our problems with the quasitriangular structure of $U_q(sl_2)$, which required us to work either at a root of unity or with formal powerseries, do not appear in the dual quasitriangular setting. We may work with $q \in k^*$ or indeed with q (or more precisely $q^{\frac{1}{2}}$) an indeterminate when we work with $SL_q(2)$ instead.

Example 8.8 *Let $q \in k^*$ have a square root. Then the Hopf algebra $SL_q(2)$ is dual quasitriangular, with $\mathcal{R}$ on the basis of generators a, b, c, d*

given by the matrix

$$\mathcal{R} = q^{-\frac{1}{2}} \begin{pmatrix} q & 0 & 0 & 1 \\ 0 & 0 & q - q^{-1} & 0 \\ 0 & 0 & 0 & 0 \\ 1 & 0 & 0 & q \end{pmatrix}.$$

Proof We will see general constructions for this much later in the course. For the moment, the bicharacter axioms specify the extension to products of the generators and one can (in principle) verify directly that this extension is compatible with the algebra relations. One may first establish it for the q-commutation relations (those of the bialgebra $M_q(2)$), giving $M_q(2)$ as a quasitriangular bialgebra, and then verify that $\mathcal{R}$ descends to the quotient $SL_q(2)$ by the q-determinant relation. Note also that the $M_q(2)$ relations are homogeneous in degree (quadratic) and hence any overall factor in the role of $q^{-\frac{1}{2}}$ can be used, while descending to $SL_q(2)$ does fix the normalisation. For example,

$$\mathcal{R}(a \otimes ad - q^{-1}bc) = \mathcal{R}(a \otimes a)\mathcal{R}(a \otimes d) = 1 = \epsilon(a) = \mathcal{R}(a \otimes 1)$$

from $\Delta a = a \otimes a + b \otimes c$ and the stated form of $\mathcal{R}$ on the generators. □

Note that we have by no means given a full proof that $\mathcal{R}$ extends (merely a fix of the normalisation); we defer that to much later in the course when we have some powerful 'R-matrix' techniques. Just to get thinking about this, however, let us note that from Lemma 8.5 we have an immediate corollary:

Corollary 8.9 *If H is a dual quasitriangular bialgebra or Hopf algebra with a matrix $t^i{}_j$ say of generators and a matrix form*

$$\Delta t^i{}_j = \sum_k t^i{}_k \otimes t^k{}_j, \quad \epsilon(t^i{}_j) = \delta^i{}_j$$

of coalgebra, then the matrix $R \in M_n \otimes M_n$ defined by

$$R^i{}_j{}^k{}_l = \mathcal{R}(t^i{}_j \otimes t^k{}_l)$$

obeys the Yang–Baxter equation in $M_n \otimes M_n \otimes M_n$. Moreover, the relations

$$\sum_{a,b} R^i{}_a{}^k{}_b t^a{}_j t^b{}_l = \sum_{a,b} t^k{}_b t^i{}_a R^a{}_j{}^b{}_l$$

hold in H for all i, j, k, l.

Proof Immediate from the lemma. The second part is a consequence of the dual quasitriangularity axioms and the form of the coproduct.   □

Later on, we will prove the converse: if R is a matrix solution of the Yang–Baxter equations then there is a dual quasitriangular bialgebra $A(R)$ with these relations.

This completes the first 1/3 of the course, which covers the basic Hopf algebra theory and the basic examples. In the next lecture we will start the second part of the course, which is the representation theory of quantum groups, leading to braided categories, knot invariants etc. The last 1/3 of the course will be the more advanced quantum group theory and elements of noncommutative geometry.

9

Braided categories

In this lecture we start a block of the course in which we study the representation theory of quantum groups and its applications. We will see that they are intimately connected with braids and knots.

We start with some abstract definitions of monoidal and braided categories. A category $\mathcal{C}$ for our purposes is just

1. A collection of *objects* $V, W, Z, U, \ldots$.

2. A specification of a set $\mathrm{Mor}(V, W)$ of *morphisms* for each V, W. The sets $\mathrm{Mor}(V, W)$, $\mathrm{Mor}(Z, U)$ are disjoint unless $V = Z$ and $W = U$.

3. A composition operation $\circ : \mathrm{Mor}(W, Z) \times \mathrm{Mor}(V, W) \to \mathrm{Mor}(V, Z)$ with properties analogous to the composition of maps (such as associativity of $\circ$ where defined).

4. Every set $\mathrm{Mor}(V, V)$ should contain an identity element id_V such that $\phi \circ \mathrm{id} = \phi$, $\mathrm{id} \circ \phi = \phi$ for any morphism for which $\circ$ is defined.

A more formal treatment is in Mac Lane's book for anyone interested. In our case all objects will be concrete sets with structure (actually vector spaces equipped with linear maps of various kinds), all morphisms will be linear maps obeying various restrictions, and all categories will be equivalent to essentially small ones (i.e. we will not digress on topos theory and other subtleties). We are primarily going to use the language of category theory to keep our thinking clear. In particular, we indicate objects as $V \in \mathcal{C}$ by an abuse of set theory notations.

A (covariant) *functor* $F : \mathcal{C} \to \mathcal{V}$ between categories specifies an object $F(V) \in \mathcal{V}$ for every object $V \in \mathcal{C}$, *and a* morphism $F(\phi) : F(V) \to F(W)$ for every morphism $\phi : V \to W$, such that $F(\phi \circ \psi) = F(\phi) \circ F(\psi)$ for morphisms. A 'contravariant functor' is similar but with $F(\phi) : F(W) \to F(V)$ and $F(\phi \circ \psi) = F(\psi) \circ F(\phi)$.

Less obvious is the notion of a *natural transformation* $\theta : F \to G$ or $\theta \in \mathrm{Nat}(F, G)$ between two functors $F, G : \mathcal{C} \to \mathcal{V}$. This means, in fact,

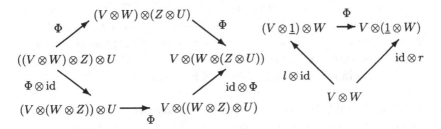

Fig. 9.1. Pentagon condition for Φ and triangle condition for compatibility with l, r.

a collection $\{\theta_V : F(V) \to G(V) \mid V \in \mathcal{C}\}$ of morphisms in $\mathcal{V}$ which are 'functorial' in the sense that

$$\theta_W \circ F(\phi) = G(\phi) \circ \theta_V, \quad \forall \phi : V \to W$$

(in the covariant case; similarly in the contravariant case). The natural transformation θ is called a 'natural isomorphism' if each θ_V is an isomorphism.

An obvious example is the category $_A\mathcal{M}$ of modules over an algebra A. The morphisms are maps commuting with the action of A. We have a lot of structure in this category: direct sums, k-linearity etc.

Definition 9.1 *A monoidal category is $(\mathcal{C}, \otimes, \underline{1}, \Phi, l, r)$, where*

1. $\mathcal{C}$ is a category.

2. $\otimes : \mathcal{C} \times \mathcal{C} \to \mathcal{C}$ is a functor.

3. A natural isomorphism $\Phi : (\ \otimes\ ) \otimes \to \otimes(\ \otimes\ )$, i.e. a collection of functorial isomorphisms

$$\Phi_{V,W,Z} : (V \otimes W) \otimes Z \cong V \otimes (W \otimes Z), \quad \forall V, W, Z \in \mathcal{C},$$

obeying the 'pentagon condition' in Figure 9.1.

4. A unit object $\underline{1}$ and associated natural isomorphisms $l : (\ ) \otimes \underline{1} \to$ id, $r : \underline{1} \otimes (\ ) \to$ id, i.e. a collection of functorial isomorphisms $l_V : V \cong V \otimes \underline{1}$, $r_V : V \cong \underline{1} \otimes V$, obeying the 'triangle condition' in Figure 9.1.

The pentagon condition equates the two ways to go $((V \otimes W) \otimes Z) \otimes U \to V \otimes (W \otimes (Z \otimes U))$ by applying Φ repeatedly. Mac Lane's coherence theorem asserts that all other consistency problems of this nature are then automatically solved as well. This means, in practice, that we can just omit brackets and write expressions such as $V \otimes W \otimes Z \otimes U$ quite freely. There will be several ways to fill in the brackets and Φ in our expressions, but all the different ways will coincide. The maps l, r associated to the unit object also take care of themselves once the consistency condition stated is satisfied. We will therefore soon suppress these maps. In fact, the category Vec of vector spaces is monoidal with the usual $\otimes$ and Φ, l, r will mostly just be inherited from this, i.e. trivial.

Proposition 9.2 *If H is a bialgebra or Hopf algebra then its module category $_H\mathcal{M}$ is monoidal with $\otimes$ defined by $h\triangleright(v \otimes w) = \sum h_{(1)}\triangleright v \otimes h_{(2)}\triangleright w$ for all $h \in H$, $v \in V$ and $w \in W$, and $\underline{1} = k$ (as explained in Lecture 2).*

Proof We have already checked that $V \otimes W$ is a representation of H if V, W are. Most of the parts of the definition of a monoidal category are inherited from Vec and easily checked. In particular, we take

$$\Phi_{V,W,Z}((v \otimes w) \otimes z) = v \otimes (w \otimes z), \quad \forall v \in V, \ w \in W, z \in Z$$

as for vector spaces. We have to check that it is a morphism in $_H\mathcal{M}$. Thus,

$$\begin{aligned}
h\triangleright((v \otimes w) \otimes z) &= h_{(1)(1)}\triangleright v \otimes h_{(1)(2)}\triangleright w \otimes h_{(2)}\triangleright z \\
&= h_{(1)}\triangleright v \otimes h_{(2)(1)}\triangleright w \otimes h_{(2)(2)}\triangleright z = h\triangleright(v \otimes (w \otimes z))
\end{aligned}$$

by coassociativity of Δ. The obvious maps l, r (as for Vec) are likewise morphisms, by the counity axioms. $\square$

Next, we write $\otimes^{\mathrm{op}}(V, W) = W \otimes V$. For a general Hopf algebra $\otimes$ and $\otimes^{\mathrm{op}}$ can be completely unrelated. For groups or Lie algebra representations, by contrast, $V \otimes W$ and $W \otimes V$ are trivially identified, corresponding to cocommutativity of the group algebra or enveloping algebra. For a quasitriangular Hopf algebra we will find that they are isomorphic, but by a nontrivial isomorphism called the 'braiding' Ψ.

Definition 9.3 A braided monoidal *category* $(\mathcal{C}, \otimes, \underline{1}, \Phi, l, r, \Psi)$ *is*
1. A monoidal category $(\mathcal{C}, \otimes, \underline{1}, \Phi, l, r)$.
2. A natural isomorphism $\Psi : \otimes \to \otimes^{\mathrm{op}}$, *i.e. a collection of functorial isomorphisms*

$$\Psi_{V,W} : V \otimes W \to W \otimes V, \quad \forall V, W \in \mathcal{C},$$

obeying the 'hexagon condition' in Figure 9.2.

The model here is the usual twist or transposition map $V \otimes W \cong W \otimes V$ for vector spaces. If we suppress Φ, then the hexagon conditions are

$$\Psi_{V \otimes W, Z} = \Psi_{V,Z} \circ \Psi_{W,Z}, \quad \Psi_{V,W \otimes Z} = \Psi_{V,Z} \circ \Psi_{V,W}$$

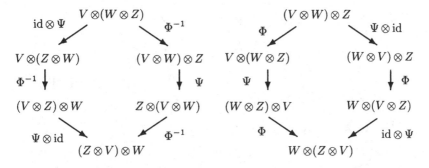

Fig. 9.2. Hexagon conditions for Ψ.

for all objects V, W, Z, i.e. transposing $V \otimes W$ past Z is the same as transposing W past Z and then V past Z, and transposing V past $W \otimes Z$ is the same as first transposing V past W and then V past Z. These are natural properties that we might expect for any reasonable 'transposition' map. From the hexagons one then deduces among other things that

$$\Psi_{V,\underline{1}} = \mathrm{id}, \quad \Psi_{\underline{1},V} = \mathrm{id}.$$

On the other hand, there is one important property of the usual transposition for vector spaces that we do *not* carry over. Namely, we do not assume that $\Psi_{V,W} = \Psi_{W,V}^{-1}$ (if this holds for all V, W then we have a 'symmetric' monoidal category). This leads to the following convenient notation for working with such braidings. We write morphisms pointing generally downwards (say) and denote tensor product by horizontal juxtaposition. Instead of a usual arrow for Ψ, Ψ^{-1} we use the shorthand

$$\Psi_{V,W} = \begin{array}{c} V \quad W \\ \diagdown\!\!\diagup \\ W \quad V \end{array} , \quad (\Psi_{W,V})^{-1} = \begin{array}{c} V \quad W \\ \diagdown\!\!\diagup \\ W \quad V \end{array}$$

to distinguish them. We denote any other morphisms as nodes on a string with the appropriate number of input and output legs. In this notation, the hexagon conditions and the functoriality of the 'braiding' Ψ appear as shown in Figure 9.3, where the doubled lines in part (a) refer to the composite objects $V \otimes W$ and $W \otimes Z$ in a convenient extension of the notation. The functoriality of Ψ is expressed in part (b) as the assertion that a morphism $\phi : V \rightarrow Z$ can be pulled through a braid crossing. Similarly for Ψ^{-1} with inverse braid crossings.

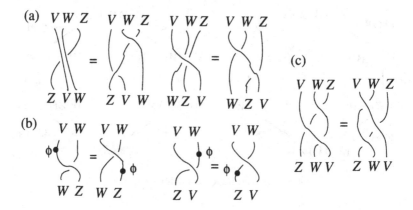

Fig. 9.3. Hexagons (a) and functoriality (b) in the diagrammatic notation. They imply the Yang–Baxter identity or braid relations (c).

The coherence theorem for braided categories (written out formally by Joyal and Street) asserts that if two composite morphisms

$$V_1 \otimes V_2 \otimes \cdots \otimes V_n \to V_{\sigma(1)} \otimes V_{\sigma(2)} \otimes \cdots V_{\sigma(n)}$$

(for some permutation σ and some bracketings on each side) that are built from $\Psi, \Psi^{-1}, \Phi, \Phi^{-1}$ correspond to the same braid, then they coincide as morphisms. For example, the two braids in Figure 9.3(c) coincide because $\Psi_{V,W}$ on the left can be pushed up over the Z line. It corresponds to the identity

$$\Psi_{V,W} \circ \Psi_{V,Z} \circ \Psi_{W,Z} = \Psi_{W,Z} \circ \Psi_{V,Z} \circ \Psi_{V,W},$$

which indeed holds by the hexagon identities and functoriality. The Artin braid group on a given number of strands is the group generated by under and over braid crossings of adjacent strands, regarded as mutually inverse, and the braid relations in Figure 9.3(c) for three strands. We will see in the next lecture that these are a version of the Yang–Baxter relations and that every quasitriangular bialgebra gives rise to a braided category as its category of modules.

We will not have space in the course to cover the case of nontrivial Φ (for example the theory of quasi-Hopf algebras), but here is a simple example. Let G be a finite group and $\phi : G \times G \times G \to k^*$ a group 3-cocycle in the sense

$$\phi(g,e,h) = 1, \quad \phi(gh,s,t)\phi(g,h,st) = \phi(g,h,s)\phi(g,hs,t)\phi(h,s,t)$$

for all $g, h, s, t \in G$. Then the category of G-graded vector spaces is a monoidal one with tensor product the usual one (and the group composition of the gradings), and

$$\Phi_{V,W,Z}((v \otimes w) \otimes z) = v \otimes (w \otimes z)\phi(|v|, |w|, |z|)$$

for all $v \in V$, $w \in W$, $z \in Z$ of homogeneous degree $|v|, |w|, |z| \in G$. A map $\mathcal{R} : G \times G \to k^*$ obeying some other conditions (a 'quasibicharacter') defines similarly a braiding Ψ in this category. For a concrete example,

$$G = \mathbb{Z}_{/2} \times \mathbb{Z}_{/2} \times \mathbb{Z}_{/2}, \quad \phi(v, w, z) = (-1)^{(v \times w) \cdot z}$$

where we consider elements of G as $\mathbb{Z}_{/2}$-valued vectors and use the vector cross and dot products. This defines the monoidal category wherein naturally live the octonions.

10

(Co)module categories. Crossed modules

We are now ready to see the point behind all the abstractions in the last lecture. One thing that we've seen is that if V is an object in a braided category then on $V \otimes V \otimes \cdots \otimes V$ (n times) we have a representation of the Artin braid group B_n on n strands. This is the free group generated by b_i, b_i^{-1} for $i = 1, \ldots, n-1$ modulo the relations

$$b_i b_j = b_j b_i, \quad \forall |i - j| \geq 2, \quad b_i b_{i+1} b_i = b_{i+1} b_i b_{i+1}$$

and the implied relations for b_i^{-1}. This is the structure of

$$b_i = |\,| \cdots |\,\mathsf{X}\,| \cdots |\,|, \quad b_i^{-1} = |\,| \cdots |\,\mathsf{X}\,| \cdots |\,|$$

(braids in the $i, i+1$ position), forming a group under composition and the obvious cancellation move. Note that the braid group projects onto the symmetric group S_n by the further relation $b_i^2 = e$ (the group identity), i.e. when b_i and b_i^{-1} coincide. The representation of B_n on $V^{\otimes n}$ is of course by sending b_i to $\Psi_{i,i+1}$ (this means $\Psi_{V,V}$ acting in the $i, i+1$ copies) and b_i^{-1} to its inverse. This is the reason that Ψ is called a 'braiding'.

On the other hand, specifying a linear representation of B_n of this type is the same thing as finding invertible $\Psi \in \text{End}(V \otimes V)$ such that

$$\Psi_{12} \Psi_{23} \Psi_{12} = \Psi_{23} \Psi_{12} \Psi_{23},$$

which is the same thing as the Yang–Baxter equation for invertible $R \in \text{End}(V \otimes V)$ when $\Psi = \tau \circ R$. Of course, a braided category has much more structure than this action of B_n, but we see that it has a lot to do with the representation theory of quasitriangular Hopf algebras.

Proposition 10.1 *Let H be a quasitriangular bialgebra or Hopf algebra.*

Then the category $_H\mathcal{M}$ of H-modules is braided, with

$$\Psi_{V,W}(v \otimes w) = \sum \mathcal{R}^{(2)} \triangleright w \otimes \mathcal{R}^{(1)} \triangleright v, \quad \forall v \in V, \ w \in W.$$

Proof We verify first that Ψ as stated is indeed a morphism, i.e. commutes with the action of H. We have

$$\begin{aligned}
\Psi_{V,W}(h \triangleright (v \otimes w)) &= \Psi_{V,W}((\Delta h) \triangleright (v \otimes w)) = \tau(\mathcal{R}(\Delta h) \triangleright (v \otimes w)) \\
&= \tau((\tau \circ \Delta h)\mathcal{R} \triangleright (v \otimes w)) = h \triangleright \Psi_{V,W}(v \otimes w)
\end{aligned}$$

for all $h \in H$, $v \in V$ and $w \in W$, as required. We used $\triangleright$ to also denote the action of $H \otimes H$ on $V \otimes W$ in the obvious way. The usual transposition map alone will not in general be a morphism: we need to apply $\mathcal{R}$ first. Thus, Ψ a morphism corresponds precisely to the quasi-cocommutativity axiom of a quasitriangular structure. Similarly, the other axiom 3 of a quasitriangular structure corresponds to the hexagon conditions. For example,

$$\begin{aligned}
\Psi_{V,W \otimes Z}(v \otimes w \otimes z) &= \tau_{23} \circ \tau_{12}(((\mathrm{id} \otimes \Delta)\mathcal{R}) \triangleright (v \otimes w \otimes z)) \\
&= \tau_{23} \circ \tau_{12}(\mathcal{R}_{13}\mathcal{R}_{12} \triangleright (v \otimes w \otimes z)) \\
&= \tau_{23} \circ \mathcal{R}_{23} \triangleright ((\tau \circ \mathcal{R} \triangleright (v \otimes w)) \otimes z) \\
&= \Psi_{V,Z} \circ \Psi_{V,W}(v \otimes w \otimes z)
\end{aligned}$$

in the compact notation where $\triangleright$ also denotes the action of $H \otimes H \otimes H$ on $V \otimes W \otimes Z$ and the numerical suffixes denote the position in a multiple tensor product. Similarly for the other half of this axiom. One can write out the proofs in a more explicit notation if preferred. Also, the form of Ψ whereby it is given by an element of $H \otimes H$ acting (followed by the usual transposition map for vector spaces) ensures functoriality. Finally, the assumption that $\mathcal{R}$ is invertible ensures that the Ψ are invertible. $\square$

This tells us that the axioms of a quasitriangular bialgebra are exactly what it takes to make the category of modules have a braiding of this form. Later on in the course we will outline the converse theorem more precisely. It should also be clear that the triangular case where $\mathcal{R}_{21} = \mathcal{R}^{-1}$ corresponds precisely to the case where Ψ gives a symmetric monoidal category rather than a truly braided one. The factorisable case is the one maximally far from this. We turn now to some elementary examples of braided categories constructed by the proposition.

Example 10.2 *Let q be a primitive n'th root of 1. The braided category* *of $k_q\mathbb{Z}_{/n}$-modules consists of $\mathbb{Z}_{/n}$-graded vector spaces with morphisms linear maps that preserve the grading. The tensor product is $|v \otimes w| = |v| + |w|$ modulo n, and the braiding is*

$$\Psi_{V,W}(v \otimes w) = q^{|v||w|} w \otimes v, \quad \forall v \in V, \ w \in W$$

of homogeneous degrees $|v|, |w|$ respectively. The category is truly braided for $n > 2$.

Proof Recall that $k_q\mathbb{Z}_{/n} = k[g]$ modulo $g^n = 1$. Hence its representations decompose as $V = \bigoplus_{a=0}^{n-1} V_a$, where g acts as $g \triangleright v = q^a v$ for all $v \in V_a$. We say that $v \in V_a$ has degree $|v| = a$. This is the grading. The coproduct is $\Delta g = g \otimes g$, so the decomposition of a tensor product representation is by adding the degrees. The formula for $\mathcal{R}$ given in Lecture 6 then immediately gives the braiding shown.   □

The $n = 2$ case is exactly the symmetric category of $\mathbb{Z}_{/2}$-graded or supervector spaces and even maps as morphisms, which is a category that occurs throughout physics and mathematics. So we identify supervector spaces as the module category (one can say 'generated by') the triangular Hopf algebra $k_q\mathbb{Z}_{/2}$. Note that this statement could not be made before the introduction of all our theory.

Example 10.3 *Let G be a finite group and let $D(G) = D(kG)$ be the quantum double of its group Hopf algebra. Its category of modules consists of objects V where*

1. V is a kG-module (a linear representation of G).

2. V is G-graded such that $|g \triangleright v| = g|v|g^{-1}$ for all $g \in G$ and v of degree $|v| \in G$.

The morphisms are maps that commute with the action of G and preserve the grading. The tensor product has the diagonal action of G and the grading $|v \otimes w| = |v||w|$ for homogeneous elements. This category is truly braided (for G nontrivial) with

$$\Psi(v \otimes w) = |v| \triangleright w \otimes v, \quad \forall v \in V, \ w \in W$$

with v of homogeneous degree.

Proof Here $D(G)$ is built in $k(G) \otimes kG$ with the product, coproduct, antipode and factorisable quasitriangular structure

$$(\delta_a \otimes g)(\delta_b \otimes h) = \delta_{g^{-1}ag,b}\delta_b \otimes gh,$$

$$\Delta(\delta_g \otimes h) = \sum_{ab=g} \delta_a \otimes h \otimes \delta_b \otimes h, \quad \epsilon(\delta_g \otimes h) = \delta_{g,e},$$

$$S(\delta_g \otimes h) = \delta_{h^{-1}g^{-1}h} \otimes h^{-1}, \quad \mathcal{R} = \sum_{g \in G} \delta_g \otimes e \otimes 1 \otimes g$$

for $a, b, g, h \in G$. Here $e \in G$ is the identity and δ_g denotes the Kronecker delta-function $\delta_g(h) = \delta_{g,h}$ for all $g, h \in G$. Clearly, $\{\delta_g \otimes h \mid g, h \in G\}$ is a basis of $D(G)$. Since kG and $k(G)$ are sub-Hopf algebras of $D(G)$, a $D(G)$-module V pulls back by restriction to a kG-module and a $k(G)$-module, as stated. Here a $k(G)$ action corresponds to a grading by $\phi \triangleright v = \phi(|v|)v$ for $\phi \in k(G)$ and $v \in V$ of degree $|v|$. The further relations of $D(G)$ then require the compatibility of these two actions. The form of the coproduct tells us the action and grading on tensor products, and finally

$$\Psi_{V,W}(v \otimes w) = \sum_{g \in G} (1 \otimes g) \triangleright w \otimes (\delta_g \otimes e) \triangleright v$$

$$= \sum_{g \in G} g \triangleright w \otimes \delta_g(|v|)v = |v| \triangleright w \otimes v$$

for all $v \in V$ and $w \in W$ with v of homogeneous degree $|v|$. $\square$

This is the category of crossed kG-modules introduced some decades ago by J.H.C. Whitehead in algebraic topology (but without knowing about the braiding). More precisely, he introduced what are called crossed G-sets were G acts on a set, with some other conditions in that context; we have a linearised version. It motivates the following general construction.

Proposition 10.4 *Let H be a Hopf algebra with invertible antipode. The braided category $^H_H\mathcal{M}$ of 'crossed H-modules' consists of objects V where*

    1. V is an H-module (i.e. $V \in {}_H\mathcal{M}$).

    2. V is an H-comodule (we write $V \in {}^H\mathcal{M}$) and

$$\sum h_{(1)}v^{(\bar{1})} \otimes h_{(2)} \triangleright v^{(\bar{2})} = \sum (h_{(1)} \triangleright v)^{(\bar{1})} h_{(2)} \otimes (h_{(1)} \triangleright v)^{(\bar{2})}, \quad \forall v \in V, \ h \in H$$

where $\triangleright$ denotes the action on V and $v \mapsto \sum v^{(\bar{1})} \otimes v^{(\bar{2})}$ the coaction. Morphisms commute with both the action and coaction. The tensor *product is that of the action and coaction. The braiding is*

$$\Psi_{V,W}(v \otimes w) = \sum v^{(\bar{1})} \triangleright w \otimes v^{(\bar{2})}, \quad \forall v \in V, \; w \in W.$$

Proof One can check that the compatibility condition holds for the tensor product object $V \otimes W$, that Ψ is a morphism and that the two hexagon conditions hold. The other details of a braided category are all trivial. Actually, we omit the proof for the following reason: when H is finite dimensional, this category is just $_{D(H)}\mathcal{M}$. Recall that the quantum double $D(H)$ contains H and $H^{*\mathrm{op}}$ as sub-Hopf algebras. So an object of its module category is an H-module and an $H^{*\mathrm{op}}$ (left) module, subject to a compatibility. But the latter is just the same thing as a right H^*-module, which is the same thing as evaluation against a left coaction of H. If $V \in {}_H^H\mathcal{M}$, the corresponding action of $D(H)$ is

$$(\phi \otimes h) \triangleright v = \langle \phi, (h \triangleright v)^{(\bar{1})} \rangle (h \triangleright v)^{(\bar{2})}, \quad v \in V, \; \phi \in H^*, \; h \in H.$$

So we would be repeating the construction of the $D(H)$ as a quasitriangular bialgebra, just with the action of $H^{*\mathrm{op}}$ replaced by a coaction of H. For example, from the $D(H)$ point of view,

$$\Psi_{V,W}(v \otimes w) = \sum_a e_a \triangleright w \otimes f^a \triangleright v$$

$$= \sum_a e_a \triangleright w \otimes \langle f^a, v^{(\bar{1})} \rangle v^{(\bar{2})} = v^{(\bar{1})} \triangleright w \otimes v^{(\bar{2})}$$

for $v \in V$ and $w \in W$, where $\{e_a\}$ is a basis of H and $\{f^a\}$ is a dual basis. It is also worth noting that for a braided category we need only H as a bialgebra and H^{cop} (with opposite coalgebra) a Hopf algebra, i.e. we need a map with the properties of S^{-1} (it is needed to define Ψ^{-1}), rather than S itself. In the finite-dimensional case we need $D(H)$ as a quasitriangular bialgebra, not actually as a Hopf algebra. However, later on when we consider dual representations, we will need S also.   □

This is the infinite-dimensional version of the module category of the quantum double (and can be used to define $D(H)$ when H is infinite dimensional). Let us note now that every Hopf algebra is itself canonically an object in $_H^H\mathcal{M}$.

Corollary 10.5 $H \in {}_H^H\mathcal{M}$ *by* Ad *the adjoint action and* Δ *the left regular coaction. Hence we have a 'canonical braiding'*

$$\Psi_{H,H} : H \otimes H \to H \otimes H, \quad \Psi_{H,H}(g \otimes h) = \sum g_{(1)} h S g_{(2)} \otimes g_{(3)}$$

for $g, h \in H$ associated to any Hopf algebra, and an action of the Artin braid group B_n on $H^{\otimes n}$ for all $n > 1$.

Proof This is elementary. We have

$$
\begin{aligned}
h_{(1)}g_{(1)} \otimes \mathrm{Ad}_{h_{(2)}}(g_{(2)}) &= h_{(1)}g_{(1)} \otimes h_{(2)}g_{(2)}Sh_{(3)} \\
&= h_{(1)}g_{(1)}(Sh_{(4)})h_{(5)} \otimes h_{(2)}g_{(2)}Sh_{(3)} \\
&= \mathrm{Ad}_{h_{(1)}}(g)_{(1)}h_{(2)} \otimes \mathrm{Ad}_{h_{(1)}}(g)_{(2)}
\end{aligned}
$$

for all $h, g \in H$, as required for the compatibility between the module and comodule. The rest is then a corollary of the proposition. $\qquad\square$

This is a nontrivial braiding even for classical Hopf algebras. If G is a finite group then $H = kG$ yields the braiding

$$
\Psi_{kG,kG} : kG \otimes kG \to kG \otimes kG, \quad \Psi_{kG,kG}(g \otimes h) = ghg^{-1} \otimes g, \quad \forall g, h \in G.
$$

For another example, we can let $\mathfrak{g}$ be a Lie algebra and take $H = U(\mathfrak{g})$. In this case the action and coaction also restrict to $V = k1 \oplus \mathfrak{g}$ as a finite-dimensional subobject of $U(\mathfrak{g})$ in ${}^{U(\mathfrak{g})}_{U(\mathfrak{g})}\mathcal{M}$. Its braiding is

$$
\Psi_{V,V} : V \otimes V \to V \otimes V, \quad \Psi_{V,V}(x \otimes y) = [x, y] \otimes 1 + y \otimes x, \quad \forall x, y \in \mathfrak{g},
$$

and $\Psi(1 \otimes x) = x \otimes 1$ etc. So we have a (nontrivial) linear representation of the braid group (or Yang–Baxter operator) associated to any finite group and any nontrivial Lie algebra.

11

q-Hecke algebras

We come now to the representations of the quantum group $U_q(sl_2)$. We let $q \in k^*$ be generic in the sense that $_q[n] = \frac{q^n - q^{-n}}{q - q^{-1}}$ are nonzero for all n, and $q, _q[n]$ have square roots, which we fix. The spin j or $2j + 1$-dimensional representations V_j of $U_q(sl_2)$ are the following, for $j \in \frac{1}{2}\mathbb{Z}_+$. As basis we take

$$V_j = \{e_m^j \mid m = -j, -j+1, \ldots, j-1, j\}$$

and the action

$$g \triangleright e_m^j = q^{2m} e_m^j, \quad E \triangleright e_m^j = q^m \sqrt{_q[j - m] \, _q[j + m + 1]} e_{m+1}^j$$

$$F \triangleright e_m^j = q^{-(m-1)} \sqrt{_q[j + m] \, _q[j - m + 1]} e_{m-1}^j.$$

It is elementary to check that this is indeed a representation for each j. In the setting over $\mathbb{C}[\![t]\!]$, recall that E, F, g are replaced by generators $X_\pm, H$ with relations which to lowest order are the relations of sl_2. The above representations to lowest order are the standard finite-dimensional irreducible unitarisable representations of sl_2. In this formal powerseries setting the algebras of $U_q(sl_2)$ and $U(sl_2)$ are actually isomorphic, which is the deeper reason for the 1-1 correspondence.

For general fields, the V_j are still a nice class of representations with properties analogous to those for sl_2. They are highest-weight representations in the sense that $E \triangleright e_j^j = 0$ and $g \triangleright e_j^j$ is a multiple of e_j^j. The other basis elements are generated by applying F. Let us also note in passing that one does not really need the square roots, they are more a matter of convention. One can also work with the $n + 1$-dimensional representations

$$V_n = \{v_p^n \mid p = 0, 1, \ldots, n\}, \quad n \in \mathbb{Z}_+$$

and action

$$g \triangleright v_p^n = q^{2p-n} v_p^n, \quad E \triangleright v_p^n = {}_q[p+1] v_{p+1}^n, \quad F \triangleright v_p^n = {}_q[n-p+1] v_{p-1}^n.$$

This is needed if one wants to work over $\mathbb{Q}(q)$, for example.

Example 11.1 *Let $q \in k^*$ be generic in the sense above. The subcategory of those $U_q(sl_2)$-modules which are isomorphic to a direct sum of the $\{V_j\}$ is a braided monoidal one. The braiding is*

$$\Psi_{V_i, V_j}(e_m^i \otimes e_n^j) = \sum_{a,b} (R^{(ij)})^a{}_m{}^b{}_n e_b^j \otimes e_a^i,$$

$$(R^{(ij)})^a{}_m{}^b{}_n = \delta_{m+n}^{a+b} q^{mb+na} \frac{(1-q^{-2})^{a-m}}{{}_q[a-m]!}$$

$$\sqrt{\frac{{}_q[i+a]! \; {}_q[i-m]! \; {}_q[j-b]! \; {}_q[j+n]!}{{}_q[i-a]! \; {}_q[i+m]! \; {}_q[j+b]! \; {}_q[j-n]!}}$$

for $a \geq m$, and zero otherwise.

Proof This is a matter of rather extensive verification. However, the conceptual reason is the following. First of all, the coproduct of $U_q(sl_2)$ determines the tensor product representation. Since both are deformations of the $U(sl_2)$ case, the tensor products are given by

$$V_i \otimes V_j \cong \bigoplus_k N_{ij}{}^k V_k$$

where the $N_{ij}{}^k$ are the same integer multiplicities as for $U(sl_2)$ (namely, if (say) $i \geq j$ then $N_{ij}{}^k = 1$ for k in the range $i-j, \ldots, i+j$ and zero outside this range). In our generic q setting one can check that the isomorphisms here are defined. Since the tensor product action comes from the coproduct, which differs from the usual one by g (which acts by a power of q on each e_m^j), this is more or less clear without the exact form of the isomorphisms. (They are in fact known quite explicitly in terms of q-hypergeometric functions and are a q-deformation of what are called in physics the Clebsch–Gordan coefficients or 3-j symbols.) Thus the category is closed under $\otimes$. The second observation is that E, F are nilpotent in the representations V_j, hence the quasitriangular structure $\mathcal{R}$ which we defined in the formal powerseries setting becomes in these representations a finite sum of terms. Since H acts by m on each e_m^j, the resulting Ψ is also defined in our setting of generic q. Once defined, it is clear that the algebraic proofs that $\mathcal{R}$ is a quasitriangular structure and hence defines a braided category can be repeated as a direct proof that Ψ as stated defines a braiding. □

We will need these formulae later only in the case $V = V_{\frac{1}{2}}$. Numbering the basis elements more conventionally as $e_1 = e^{\frac{1}{2}}_{-\frac{1}{2}}$, $e_2 = e^{\frac{1}{2}}_{\frac{1}{2}}$, the matrix $R = R^{(\frac{1}{2}\frac{1}{2})}$ is

$$
R = q^{-\frac{1}{2}}
\begin{pmatrix}
q & 0 & 0 & 0 \\
0 & 1 & q - q^{-1} & 0 \\
0 & 0 & 1 & 0 \\
0 & 0 & 0 & q
\end{pmatrix}
\in \mathrm{End}(V \otimes V),
$$

where $V \otimes V$ is taken with basis $e_1 \otimes e_1, e_1 \otimes e_2, e_2 \otimes e_1, e_2 \otimes e_2$. This is called the standard su_2-type solution of the Yang–Baxter equations.

We obviously have a dual theory for dual quasitriangular Hopf algebras. We work with right comodules (for a change). Clearly, the category of right comodules over a bialgebra or Hopf algebra is monoidal with the tensor product coaction

$$
\beta_{V \otimes W}(v \otimes w) = \sum v^{(\bar{1})} \otimes w^{(\bar{1})} \otimes v^{(\bar{2})}w^{(\bar{2})}, \quad \forall v \in V, \; w \in W
$$

where $\beta_V(v) = \sum v^{(\bar{1})} \otimes v^{(\bar{2})}$ and $\beta_W(w) = \sum w^{(\bar{1})} \otimes w^{(\bar{2})}$ are the right coactions $V \to V \otimes H$ and $W \to W \otimes H$. In the dual quasitriangular case, clearly

$$
\Psi_{V,W}(v \otimes w) = \sum w^{(\bar{1})} \otimes v^{(\bar{1})} \mathcal{R}(v^{(\bar{2})} \otimes w^{(\bar{2})}), \quad \forall v \in V, \; w \in W
$$

makes the category of comodules $\mathcal{M}^H$ braided. We leave the proofs as an exercise – the theory is the arrows-reversed version of the theory we have already given for quasitriangular Hopf algebras.

Example 11.2 *Let $q \in k^*$. The dual quasitriangular Hopf algebra $k_q\mathbb{Z} = k[g, g^{-1}]$ with structure $\mathcal{R}(g^n \otimes g^m) = q^{nm}$ has as its category of comodules $\mathbb{Z}$-graded vector spaces, and the braiding $\Psi_{V,W}(v \otimes w) = q^{|v||w|}w \otimes v$ on elements of homogeneous degree $|v|, |w| \in \mathbb{Z}$.*

Proof Elementary. □

Similarly, we have seen that $SL_q(2)$ is dual quasitriangular. We have also seen that it coacts on the quantum plane $\mathbb{A}_q^2$. This coaction (say from the right) restricts to the 2-dimensional subspace V spanned by the generators $e_1 = x$ and $e_2 = y$. The induced braiding Ψ has the form

$$\Psi_{V,V}(e_i \otimes e_j) = \sum_{a,b} e_b \otimes e_a R^a{}_i{}^b{}_j$$

where R is the su_2 solution of the Yang–Baxter equations as above. This is an alternative way to work with the braided category corresponding to $U_q(sl_2)$, in the comodule rather than module formalism. The construction is quite general.

Corollary 11.3 *Let $R \in M_n \otimes M_n$ be an invertible matrix solution of the Yang–Baxter equations and $A(R)$ the dual quasitriangular bialgebra of matrix type. Then $\mathcal{C}(R) = \mathcal{M}^{A(R)}$ is a braided category canonically associated to R.*

Among the objects in $\mathcal{C}(R)$ is the n-dimensional vector space V with basis $\{e_i\}$ and coaction

$$\beta_V(e_i) = \sum_a e_a \otimes t^a{}_i$$

where $t^i{}_j$ are the matrix generators of $A(R)$. Its braiding

$$\Psi_{V,V}(e_i \otimes e_j) = \sum e_b \otimes e_a R^a{}_i{}^b{}_j$$

recovers the original R, so this comodule V is called the fundamental comodule of $A(R)$. Direct sums, tensor products, quotients and direct summands etc. of V generate other objects of $\mathcal{C}(R)$.

Finally, we return to the representations of B_n defined by these various braided categories and objects in them. We limit our remarks to the k-linear setting where the braided category consists of vector spaces, has direct sums, etc. For finite-dimensional objects the representation of B_n is obviously not faithful since B_n is infinite.

Definition 11.4 *We extend the representation*

$$\rho(b_i) = \Psi_{i,i+1} \in \mathrm{End}(V^{\otimes n}), \quad n > 2$$

to the group algebra kB_n, which therefore factors through the associated Hecke algebra defined as kB_n modulo the ideal generated by $p(b_i)$, where p is the minimal polynomial of $\Psi_{V,V}$.

For example, let $\mathfrak{g}$ be a Lie algebra and Ψ the canonical braiding on $V = k1 \oplus \mathfrak{g}$ in the last lecture. When the characteristic of k is not 2, one finds that Ψ has minimal polynomial

$$(\Psi^2 - \mathrm{id})(\Psi + \mathrm{id}) = 0.$$

Actually, one can push this backwards and find that for a vector space $\mathfrak{g}$ and linear map [,], an operator Ψ of the form given obeys the braid relations and has the above minimal polynomial *iff* [,] obeys the Jacobi and antisymmetry conditions, i.e. this is equivalent to the axioms of a Lie algebra. The associated Hecke algebra is kB_n modulo the relations

$$(b_i^2 - 1)(b_i + 1) = 0.$$

Meanwhile, for V the 2-dimensional representation of $U_q(sl_2)$ and the braiding above taken (conventionally) without the $q^{-\frac{1}{2}}$ factor, one has the minimal polynomial

$$(\Psi - q)(\Psi + q^{-1}) = 0$$

and the associated Hecke algebra $\mathcal{H}_{n,q}$ is kB_n modulo the relations

$$b_i^2 = (q - q^{-1})b_i + 1.$$

When $q = 1$ we have clearly the group algebra of the symmetric group, i.e. $\mathcal{H}_{n,1} = kS_n$. Thus, one should think of $\mathcal{H}_{n,q}$ as a deformation of the symmetric group. As such, it was extensively studied in number theory (in the context of q-special functions) long before quantum groups.

Moreover, because Ψ is a morphism, we know that ρ factors through

$$\mathcal{H}_{n,q} \to \mathrm{End}_{U_q(sl_2)}(V^{\otimes n})$$

(the endomorphisms commuting with the action of $U_q(sl_2)$ on $V^{\otimes n}$). In the setting over $\mathbb{C}[[t]]$ one can show that the image of $\mathcal{H}_{n,q}$ is the commutant in $\mathrm{End}(V^{\otimes n})$ of the image of $U_q(sl_2)$. This is a q-deformation of classical Schur–Weyl duality. Similar constructions and features hold quite generally for many quantum groups.

This is all we will say for the moment on these topics. Our motivation was the tensor product of group representations, which theory is now extended to quantum group representations. On the other hand, for any group representation V we also have a dual representation V^*. We have a similar situation for any Hopf algebra (an easy exercise): if V is a module then so is V^* with

$$(h \triangleright \phi)(v) = \phi((Sh) \triangleright v), \quad \forall h \in H, \ \phi \in V^*, \ v \in V.$$

We start now the study of the abstract properties of such dualisation.

Definition 11.5 *In a monoidal category, an object V is called* rigid *(or 'finite type') if there exists an object V^* and morphisms* $\mathrm{ev}_V : V^* \otimes V \to \underline{1}$, $\mathrm{coev}_V : \underline{1} \to V \otimes V^*$ *such that*

$$(\mathrm{id} \otimes \mathrm{ev}_V) \circ (\mathrm{coev}_V \otimes \mathrm{id}) = \mathrm{id}_V, \quad (\mathrm{ev}_V \otimes \mathrm{id}) \circ (\mathrm{id} \otimes \mathrm{coev}_V) = \mathrm{id}_{V^*}.$$

If V, W are rigid and $\phi : V \to W$ is a morphism, then

$$\phi^* = (\mathrm{ev}_W \otimes \mathrm{id}) \circ (\mathrm{id} \otimes \phi \otimes \mathrm{id}) \circ (\mathrm{id} \otimes \mathrm{coev}_V) : W^* \to V^*$$

is the dual or adjoint morphism.

We suppress the isomorphisms l, r, Φ as usual. It is easy to see that if a dual object $(V^*, \mathrm{ev}, \mathrm{coev})$ does exist then it is unique up to an isomorphism. Thus, if $(V^{*\prime}, \mathrm{ev}'_V, \mathrm{coev}'_V)$ is also a dual for V, then one can define a morphism $\theta : V^{*\prime} \to V^*$ and its inverse by

$$\theta = (\mathrm{ev}'_V \otimes \mathrm{id}) \circ (\mathrm{id} \otimes \mathrm{coev}_V), \quad \theta^{-1} = (\mathrm{ev}_V \otimes \mathrm{id}) \circ (\mathrm{id} \otimes \mathrm{coev}'_V),$$

and we easily see that

$$\mathrm{ev}'_V = \mathrm{ev}_V \circ (\theta \otimes \mathrm{id}), \quad \mathrm{coev}'_V = (\mathrm{id} \otimes \theta^{-1}) \circ \mathrm{coev}_V.$$

If every object in the category has a dual, then we say that $\mathcal{C}$ is a 'rigid monoidal' category.

Proposition 11.6 *The category of finite-dimensional modules over a Hopf algebra is rigid. Similarly for the category of finite-dimensional comodules.*

Proof Elementary. We take the same form as for the category of finite-dimensional vector spaces, namely

$$\mathrm{ev}_V(\phi \otimes v) = \phi(v), \quad \mathrm{coev}_V(\lambda) = \lambda \sum_a e_a \otimes f^a,$$

for all $v \in V$, $\phi \in V^*$ and $\lambda \in k$, where $\{e_a\}$ is a basis of V and $\{f^a\}$ is a dual basis. It is then elementary to check that these maps are morphisms when the action of H on V^* is the one given above. Similarly for coactions. $\qquad\square$

12

Rigid objects. Dual representations.
Quantum dimension

In this lecture we begin to do 'linear algebra' in a braided category. At the end of the last lecture we defined what we mean by a dual object V^* of an object V, actually in a monoidal category. In this lecture we use a diagrammatic notation for the associated maps $\mathrm{ev}_V : V^* \otimes V \to \underline{1}$ and $\mathrm{coev}_V : \underline{1} \to V \otimes V^*$, namely

$$\mathrm{ev}_V = \underset{\smile}{\overset{V^* \quad V}{}} , \qquad \mathrm{coev}_V = \underset{V \quad V^*}{\overset{\frown}{}}.$$

As in Lecture 9 we write all morphisms flowing generally downwards, denote $\otimes$ by juxtaposition and the unit object $\underline{1}$ by omission. The axioms given in Lecture 11 then appear as the 'bend-straightening axioms' in Figure 12.1(a). The adjoint morphism ϕ^* is shown in part (b). The remaining part (c) shows that if V, W have such duals then $V \otimes W$ does also, with

$$(V \otimes W)^* = W^* \otimes V^*,$$

$$\mathrm{ev}_{V \otimes W} = \mathrm{ev}_W \circ \mathrm{ev}_V, \qquad \mathrm{coev}_{V \otimes W} = \mathrm{coev}_W \circ \mathrm{coev}_V,$$

where unnecessary identity maps are suppressed. We take this as the chosen dual of $V \otimes W$ in what follows. In a similar way, if V^* is rigid it is natural to choose its dual so that $\mathrm{ev}_{V^*} = (\mathrm{coev}_V)^*$ and $\mathrm{coev}_{V^*} = (\mathrm{ev}_V)^*$.

Let us now ask the question, when is $V \cong V^{**}$? If every object in our monoidal category is rigid (one says that the category is rigid) then clearly $*$ is a contravariant functor, so the question is, is the functor $*^2$ naturally isomorphic to the identity functor? Actually, when one looks at the proof of this for finite-dimensional vector spaces, one finds a hidden transposition. So in our case we will need to suppose a braiding.

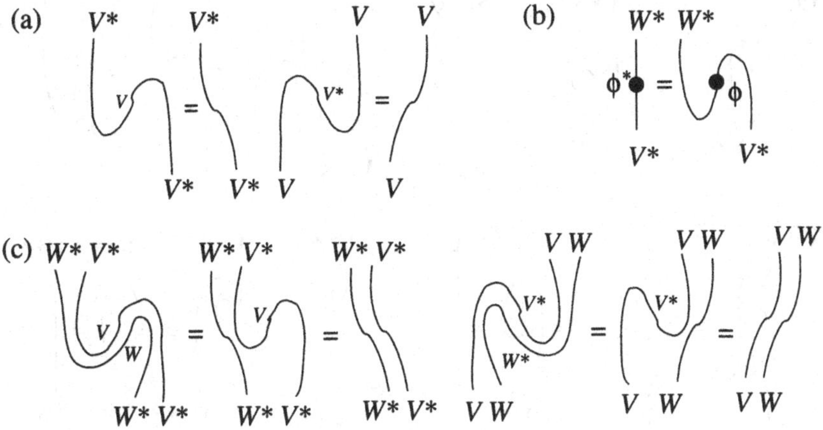

Fig. 12.1. Dual objects V^* and adjoint morphisms ϕ^*.

Proposition 12.1 *In a rigid braided monoidal category there are natural isomorphisms* $u, v^{-1} : \mathrm{id} \to *^2$, *i.e. collections of functorial isomorphisms*

$$u_V = (\mathrm{ev}_V \otimes \mathrm{id}) \circ (\Psi_{V,V^*} \otimes \mathrm{id}) \circ (\mathrm{id} \otimes \mathrm{coev}_{V^*}),$$

$$u_V^{-1} = (\mathrm{id} \otimes \mathrm{ev}_{V^*}) \circ (\Psi_{V^{**},V} \otimes \mathrm{id}) \circ (\mathrm{id} \otimes \mathrm{coev}_V),$$

$$v_V = (\mathrm{ev}_{V^*} \otimes \mathrm{id}) \circ (\mathrm{id} \otimes \Psi_{V,V^*}) \circ (\mathrm{id} \otimes \mathrm{coev}_V),$$

$$v_V^{-1} = (\mathrm{ev}_V \otimes \mathrm{id}) \circ (\mathrm{id} \otimes \Psi_{V^{**},V}) \circ (\mathrm{coev}_{V^*} \otimes \mathrm{id}),$$

obeying

$$u_{V \otimes W} = \Psi_{V,W}^{-1} \circ \Psi_{W,V}^{-1} \circ (u_V \otimes u_W),$$

$$v_{V \otimes W} = \Psi_{V,W}^{-1} \circ \Psi_{W,V}^{-1} \circ (v_V \otimes v_W).$$

Moreover,

$$(\phi^*)^* = u_W \circ \phi \circ u_V^{-1} = v_W^{-1} \circ \phi \circ v_V, \quad \forall \phi : V \to W.$$

Proof This is done diagrammatically in Figure 12.2. Our notation for ev and coev combines with our previous coherence theorem; we can slide morphisms through braid crossings, and in the case of ⌢ and

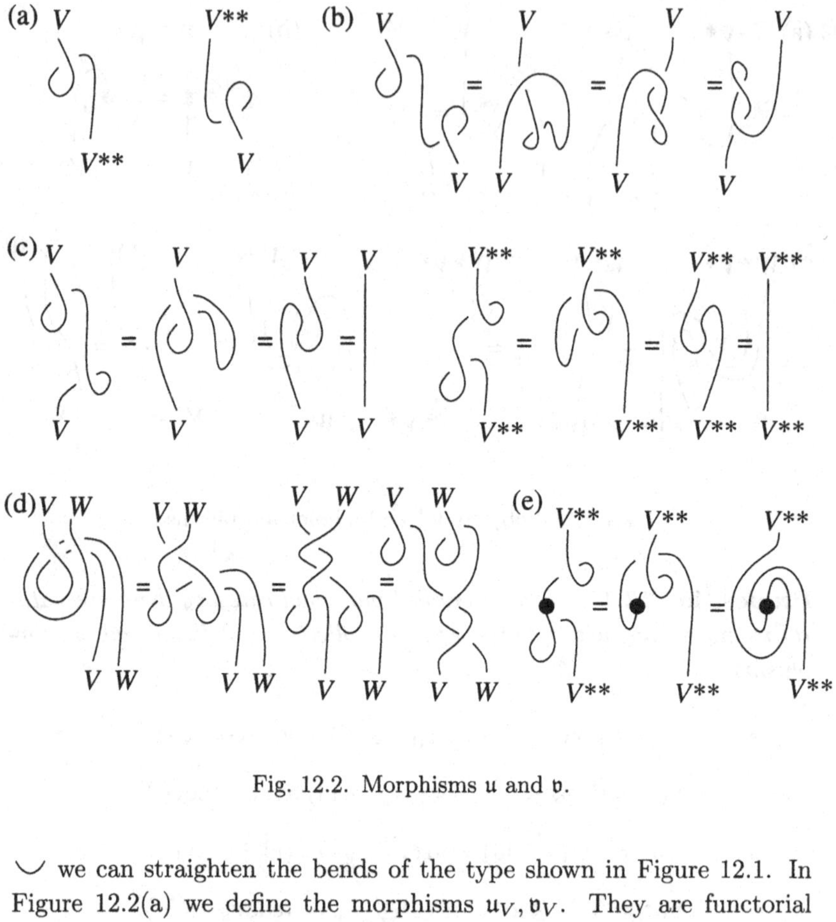

Fig. 12.2. Morphisms $\mathfrak{u}$ and $\mathfrak{v}$.

$\smile$ we can straighten the bends of the type shown in Figure 12.1. In Figure 12.2(a) we define the morphisms $\mathfrak{u}_V, \mathfrak{v}_V$. They are functorial because any other morphism or node ϕ on the string could be pulled through by functoriality of Ψ and elementary properties of ev, coev. Part (c) checks that $\mathfrak{u}$ is indeed invertible, the lower twist on the left being $\mathfrak{u}^{-1}$. The proof for $\mathfrak{v}^{-1}$ is analogous. Part (d) examines how $\mathfrak{u}$ behaves on a tensor product. Part (e) computes $\mathfrak{u} \circ \phi \circ \mathfrak{u}^{-1}$ and finds ϕ^{**} according to the definition of adjoint morphisms in Figure 12.1. Part (b) computes $\mathfrak{v} \circ \mathfrak{u}$ for later use. $\square$

Now we can do most things in linear algebra. For example, the categorical 'braided dimension' of the object V in a braided category, and the categorical 'braided trace' of an endomorphism $\phi : V \to V$, are

$$\underline{\dim}(V) = \mathrm{ev}_V \circ \Psi_{V,V^*} \circ \mathrm{coev}_V, \quad \underline{\mathrm{Tr}}_V(\phi) = \mathrm{ev}_V \circ \Psi_{V,V^*} \circ (\phi \otimes \mathrm{id}) \circ \mathrm{coev}_V$$

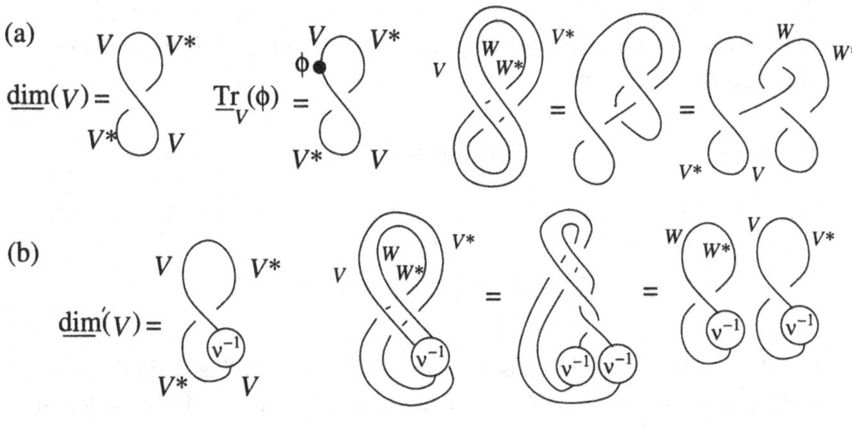

Fig. 12.3. $\underline{\dim}$ and $\underline{\mathrm{Tr}}$, nonmultiplicativity of $\underline{\dim}$ and multiplicativity of $\underline{\dim}'$.

as morphisms $\underline{1} \to \underline{1}$. They transcribe diagrammatically as shown in Figure 12.3(a). The dimension is the trace of the identity morphism. Not everything in the garden is rosy, however. For example,

$$\underline{\dim}(V \otimes W) \neq \underline{\dim}(V)\underline{\dim}(W),$$

in the general case where $\Psi^2 \neq \mathrm{id}$, with a similar problem for $\underline{\mathrm{Tr}}$. This also shown in Figure 12.3(a). To solve this problem, one has to assume more about the category.

Definition 12.2 *A braided category is called* ribbon *(or 'tortile') if the natural transformation* $\mathfrak{v} \circ \mathfrak{u}$ *has a square root natural isomorphism* $\nu : \mathrm{id} \to \mathrm{id}$ *(id the identity functor) characterised by a collection of functorial isomorphisms obeying*

$$\nu_V^2 = \mathfrak{v}_V \circ \mathfrak{u}_V, \quad \nu_{V \otimes W} = \Psi_{V,W}^{-1} \circ \Psi_{W,V}^{-1} \circ (\nu_V \otimes \nu_W),$$

$$\nu_{\underline{1}} = \mathrm{id}, \quad \nu_{V^*} = (\nu_V)^*.$$

These conditions are not independent (for example, one can conclude the first from the latter three). In this case, one can restore multiplicativity by using a modified notion $\underline{\dim}'$ of dimension, as shown in Figure 12.3(b).

Proposition 12.3 *If H is a quasitriangular Hopf algebra, then the* *functorial isomorphisms* $\mathsf{u}_V, \mathfrak{v}_V$ *in the rigid braided category of finite-dimensional H-modules are given by the action of* $\mathsf{u}, \mathfrak{v}$ *on each object V. For a ribbon quasitriangular Hopf algebra, the category is ribbon with* ν_V *given by the action of* ν.

Proof This is an exercise from the definitions above. Thus,

$$\mathsf{u}_V(v) = \sum_a (\mathrm{ev} \otimes \mathrm{id}) \circ \Psi(v \otimes f^a) \otimes E_a = \sum_a (\mathcal{R}^{(2)} \triangleright f^a)(\mathcal{R}^{(1)} \triangleright v) \otimes E_a$$

$$= \sum_a f^a((S\mathcal{R}^{(2)})\mathcal{R}^{(1)} \triangleright v) \otimes E_a = \mathsf{u} \triangleright v$$

for all $v \in V$. Here $\{f^a\}$ is a basis of V and $\{E_a\}$ is a dual basis of V^{**}. The result lies in V^{**}. The computation for $\mathfrak{v}$ is strictly analogous. Hence, if $\mathfrak{v}\mathsf{u}$ has a square root (the ribbon case), we can apply it and define $\nu_V(v) = \nu \triangleright v$ in the same way. That it obeys the condition for $\nu_{V \otimes W}$ follows from $\Delta\nu = Q^{-1}(\nu \otimes \nu)$. That it obeys the condition for ν_{V^*} corresponds likewise to $S\nu = \nu$.   □

Corollary 12.4 *Let H be a quasitriangular Hopf algebra and let V be a finite-dimensional representation.*

$$\underline{\dim}(V) = \mathrm{Tr}(\mathrm{id}) = \mathrm{Tr}(\mathsf{u}) = \underline{\dim}(V^*), \quad \underline{\mathrm{Tr}}_V(\phi) = \mathrm{Tr}(\mathsf{u} \circ \phi),$$

where $\mathsf{u}, \mathfrak{v} \in H$ *are evaluated in the representation V. The trace works for an endomorphism* $\phi : V \to V$. *In particular, the multiplicative dimension in the ribbon case is*

$$\underline{\dim}'(V) = \underline{\mathrm{Tr}}_V(\nu_V^{-1}) = \mathrm{Tr}(\nu^{-1}\mathsf{u}) = \underline{\dim}'(V^*).$$

Proof This is immediate by comparing the definition in Figure 12.3 and the definition of u_V in Figure 12.2, and then using the above proposition. It is also a nice exercise to compute directly from Figure 12.3 and the form of the braiding. Thus,

$$\underline{\dim}(V) = \sum_a \mathrm{ev} \circ \Psi(e_a \otimes f^a) = \sum_a (\mathcal{R}^{(2)} \triangleright f^a)(\mathcal{R}^{(1)} \triangleright e_a) = \sum_a f^a(\mathsf{u} \triangleright e_a),$$

where $\{e_a\}$ is a basis of V and $\{f^a\}$ is a dual basis. Similarly for the other cases.   □

One can compute this for some of our basic braided categories:

(i) For finite-dimensional $k_q \mathbb{Z}_{/n}$-modules

$$\underline{\dim}(V) = \sum_{a=0}^{n-1} q^{-a^2} \dim(V_a), \quad V = \bigoplus_a V_a.$$

This precisely generalises the usual superdimension $\dim(V_0) - \dim(V_1)$ of a supervector space $V = V_0 \oplus V_1$ (V_0 even part and V_1 odd part).

(ii) For a finite-dimensional Hopf algebra H,

$$\underline{\dim}(H) = \mathrm{Tr}S^2$$

as an object in the braided category ${}^H_H\mathcal{M}$ (the canonical braiding on any Hopf algebra as explained in Lecture 10). The number $\mathrm{Tr}S^2$ is indeed an important invariant of any finite-dimensional Hopf algebra. We see how it arises here very naturally as the categorical or 'quantum' dimension.

(iv) For the category of finite-dimensional highest-weight type $U_q(sl_2)$-modules,

$$\underline{\dim}(V_j) = q^{-2j(j+1)} {}_q[2j+1], \quad \underline{\dim}'(V_j) = {}_q[2j+1].$$

We see how the 'q-integers' are the natural 'quantum dimensions' of the corresponding representations.

In addition, every quasitriangular Hopf algebra acts on itself by the left regular representation, so when H is quasitriangular, $H \in {}_H\mathcal{M}$ and as such has a natural 'quantum order'

$$|H| \equiv \underline{\dim}(H) = \mathrm{Tr}(\mathsf{u}),$$

with u acting by multiplication on H. The usual dimension of kG is of course $|G|$. For examples,

(i) $|k_q\mathbb{Z}_{/n}| = \sum_{a=0}^{n-1} q^{-a^2}$ a theta-function on $\mathbb{Z}_{/n}$.

(ii) $|D(H)| = \mathrm{Tr}S^2$ for any finite-dimensional Hopf algebra H. This is a (slightly long) computation from our definitions.

(iii) For a further (informal) example, we consider

$$\bigoplus_j (2j+1)V_j$$

as a model for $U_q(sl_2)$ acting on itself. Our motivation is the Peter–Weyl theorem for the decomposition of the left regular representation of the compact group SU_2, which has these multiplicities $2j + 1$. Hence we define

$$|U_q(sl_2)| \equiv \sum_j (2j + 1)\underline{\dim}(V_j) = \frac{\sum_{n\in\mathbb{Z}} q^{-\frac{1}{2}n^2}}{1 - q^{-2}}.$$

Depending on the field (for example over $\mathbb{C}$ and q in a suitable region) this actually converges. $|U_q(sl_2)|$ diverges at $q = 1$, however, because the usual dimension of the regular representation (in some sense the number of points in SU_2) is infinite. But we see that this infinity becomes, by q-deformation, a pole $(1 - q^{-2})^{-1}$. It is a general feature of q-deformation that certain natural infinities become poles of this type, a phenomenon called 'q-regularisation'. These quantum order functions associated to the q-deformations of enveloping algebras have marvellous number-theoretic properties.

13

Knot invariants

We are now ready to have some fun doing knot theory. Since this is an algebra course, we will give only the most rudimentary introduction to that before some more algebraic theorems. At least it should be more or less clear what we mean by a knot – some kind of embedding of S^1 in $\mathbb{R}^3$. Two knots are equivalent or *isotopic* if they can be deformed one to the other in $\mathbb{R}^3$, i.e. there is some kind of mapping $[0, 1] \times \mathbb{R}^3 \to \mathbb{R}^3$ which is a homeomorphism of $\mathbb{R}^3$ for each $t \in [0, 1]$, is the identity at $t = 0$ and sends one knot to the other at $t = 1$. More precisely, in our case the natural setting is piecewise linear, so by a knot we mean more precisely a union of line segments closing up and non-self-intersecting in the obvious way. By isotopy we mean via a piecewise-linear map. (However, we will continue to draw curves anyway for practical reasons.) A link means a disjoint union of knots.

In fact, we will not work directly with knots and links but with their plane projections onto $\mathbb{R}^2$ in a generic direction (so that crossings in the projection are transversal and distinct from edges and from each other). This is just how we draw them on the blackboard. For example, the trefoil knot,

The first question to ask is, when do two such planar link diagrams correspond to the same actual link? This question was solved by Reidemeister who showed that two link diagrams correspond to the same link *iff* they are related by sequences of the following moves. First of all, there is a notion of general isotopy between link diagrams such as straightening double-bends, bending out arcs etc. These Reidemeister

'(0)' moves are the more obvious deformations and we suppress them (take them for granted) in what follows. Apart from these, the moves are

(1) The identifications

(2) Braid and inverse braid cancellation

(3) The braid relations

Equivalence under these moves is called 'ambient isotopy' of link diagrams. By move we mean taking one link diagram and making one of the above changes to a segment of it.

The simplest example of something that is almost invariant under these moves is the *writhe*. First of all, all our links are assumed to be oriented. This is the labelling of every arc by an arrow such that at each crossing we have one of the two possible orientations

There will be two possible orientations for each knot (or each component of a link) corresponding to flowing one way or the other along it. At each crossing p we write $\epsilon = +1$ for the first kind of crossing and $\epsilon = -1$ for the second. Then the writhe is

$$w(L) = \sum_p \epsilon(p).$$

The reader can check that it is invariant under moves (2) and (3). Moreover,

$$w(\,\bigcirc\,) = 1 + w(\downarrow) = w(\,\bigcirc\,)$$

in the sense that two links differing by replacing $\downarrow$ by $\bigcirc$ will differ in their writhe by adding 1, etc.

In fact, quantum group invariants do not usually respect the (1) move either. To deal with this one says that two link diagrams are 'regularly isotopic' if they can be related by moves (2),(3). However, given an invariant of regular isotopy one can see how it changes under the move (1) and often fix it up to be an ambient isotopy invariant by combining it in a suitable way with the writhe. So the first thing that we will head for is an invariant of regular isotopy.

Now, we have already been using these moves (2),(3) in the last lectures, in our diagrammatic notation for braided categories. Therefore, any rigid object V in a braided category provides a partially defined invariant of regular isotopy as follows: read the link diagram as a morphism $\underline{1} \to \underline{1}$. Suppose that the arcs we come to reading from the top down are $\frown$ and consider them as coev$_V$. Moving down, we apply the relevant braiding Ψ of the category for each crossing and we apply ev$_V$ when we come to an arc $\smile$. For example, the quantum dimension $\underline{\dim}(V)$ is the regular isotopy invariant associated to the figure-of-eight regular isotopy class. The invariant is only partially defined – for example, the trivial knot $\bigcirc$ does not have its arcs oriented correctly to be read as a morphism like this. On the other hand, whenever the orientation for an arc is wrong we can make it correct by a (1) move, so all link diagrams are related by (1) moves to ones which can be read as morphisms. Therefore this problem can often be fixed up at the same time as we fix the invariant to obtain an invariant of ambient isotopy.

A somewhat more systematic approach is possible in a ribbon braided category. Thus, returning to algebraic considerations, let us note that the kind of duals V^* that we have been looking at are with ev$_V$: $V^* \otimes V \to \underline{1}$. Let us call them 'left duals'. We could equally well have defined the notion of *right dual* by requiring morphisms

$$\overline{\text{ev}}_V : V \otimes V^* \to \underline{1}, \quad \overline{\text{coev}}_V : \underline{1} \to V^* \otimes V$$

obeying the mirror image of the axioms of a left dual.

Proposition 13.1 *If V has a left dual V^* (is rigid) in a ribbon braided category then V^* is also a right dual, with*

$$\overline{\text{ev}}_V = \text{ev}_V \circ (\text{id} \otimes \nu_V^{-1}) \circ \Psi_{V,V^*}, \quad \overline{\text{coev}}_V = \Psi_{V,V^*} \circ (\nu_V^{-1} \otimes \text{id}) \circ \text{coev}_V.$$

Proof We can add these $\overline{\text{ev}}, \overline{\text{coev}}$ to our diagrammatic notation, as follows. First of all, we use thick lines in our new diagrammatic notation

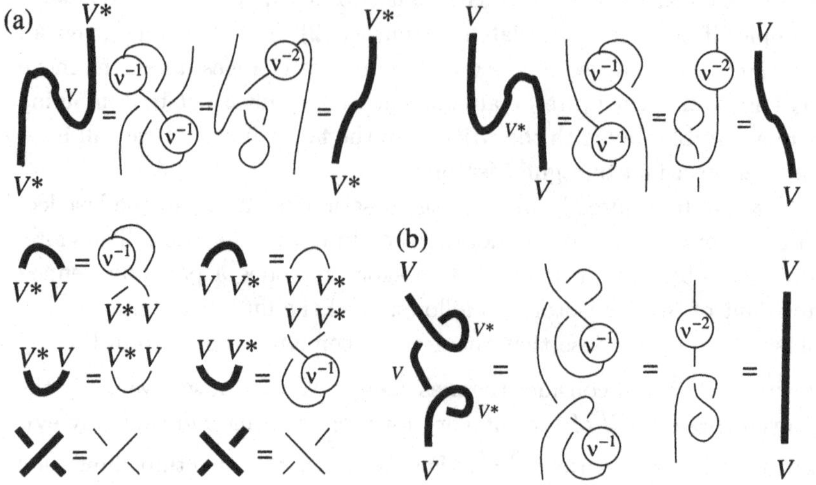

Fig. 13.1. (a) right duals and (b) an invariant of framed knots in a ribbon braided category.

to keep it clear from the old one. We write $\mathrm{ev}_V = \smile$, $\mathrm{coev}_V = \frown$ as before, but we also write $\overline{\mathrm{ev}}_V = \smile$, $\overline{\mathrm{coev}}_V = \frown$ – the two sets cannot be confused since the labelling by V, V^* is different. Figure 13.1(a) checks that the right-handed bend-straightening axioms hold for the thickened lines by expressing their definitions in terms of the unthickened old notation. The definitions in diagrammatic form are collected on the lower left of the figure. In the proof, we use the computation of $\mathfrak{v} \circ \mathfrak{u}$ in Figure 12.2(b), and then use the definition of ν as its square root. $\square$

   In this case we can obtain a genuine topological invariant, although not exactly of links but of *framed links*. By definition, a framing is (a topological equivalence class of) the choice of section of the unit normal bundle to each component knot. It is also possible to visualise a framed knot as a ribbon (of sufficiently thin width): one edge of the ribbon is the knot and the other edge is the knot displaced by a finite but sufficiently small amount along the normal vector. More precisely, there is a piecewise-linear version of these ideas which we use. On the other hand, when a link is drawn on a piece of paper with under- and over-crossings, there is a canonical framing of the link diagram, the so-called 'blackboard framing', in which the normal vector at each point comes orthogonally out of the page. Moreover, any framed link can be represented by one with the blackboard framing. We make use of this and write framed knots with the blackboard framing assumed. Then the natural topological equivalence or ambient isotopy between framed link diagrams does not involve the first Reidemeister move but instead something weaker, namely

$$(1)' \quad \text{〇} = \text{〇}$$

for pieces of a framed link with the blackboard framing. One can easily check this out with a strip of paper. We will write blackboard-framed knots with thick lines, to remind us of the framing.

Moreover, we see that in a ribbon braided category, exactly this move $(1)'$ holds. This is shown in Figure 13.1(b) using again the computation of $\mathfrak{v} \circ \mathfrak{u}$ in Figure 12.2(b). Thus we arrive at the following invariant of blackboard-framed knots and links associated to any rigid object of a ribbon braided category as follows: read it as a morphism $\underline{1} \to \underline{1}$. Thus, start at the top and label the sides of each arc as either coev if the orientation is anticlockwise (as above) or $\overline{\text{coev}}$ if the orientation is clockwise. Interpret crossings as before and interpret the final arcs as ev in the clockwise direction (as before) and $\overline{\text{ev}}$ in the anticlockwise direction. This prescription ends up with downward-labelled arcs corresponding to the identity morphism $V \to V$ and upward-labelled arcs corresponding to the identity morphism $V^* \to V^*$ [this has an interesting interpretation in physics as antiparticles being the same as particles moving backwards in time]. By our results above we see that the resulting morphism $\underline{1} \to \underline{1}$ is an invariant of ambient isotopy of framed links. Depending on its particular form, one can often extend it to an invariant of usual ambient isotopy, for example by combining with the writhe.

We will sketch now some general examples of the resulting invariant. In our concrete setting $\underline{1} = k$ so a morphism $\underline{1} \to \underline{1}$ is just an element of k.

(i) When G is a finite group, we have seen that $\mathbb{C}G \in {}^{CG}_{CG}\mathcal{M}$, the (ribbon) braided category of crossed modules (or modules over $D(G)$). The resulting invariant is of a link L is

$$\#\mathrm{Hom}(\pi_1(\overline{L}), G)$$

the number of group homomorphisms. Here $\pi_1(\overline{L})$ is the fundamental

group of the complement of the link L. I.e. we recover information about
a classical invariant, albeit a bit indirectly as the group homomorphisms
from it to G.

(ii) When V is the 2-dimensional representation of $U_q(sl_2)$ one obtains

$$J(L)(q)$$

where $J(L)$ is a version of the usual Jones polynomial $V(L)(t)$ (a polynomial in $t^{\frac{1}{2}}, t^{-\frac{1}{2}}$) associated to a link. In our conventions the Jones
polynomial is characterised by its value on the circle (a normalisation)
and

$$q^2 J_{\diagup\diagdown} - q^{-2} J_{\diagdown\diagup} = (q - q^{-1})J_{\downarrow\,\downarrow},$$

i.e. has this relationship for its values on three links differing by having
the corresponding crossing or $\downarrow\,\downarrow$ in one place. Such 'skein relations'
can be used to characterise link invariants. By contrast, the braiding on
the 2-dimensional representation of $U_q(sl_2)$ in Lecture 11 obeyed (after
multiplying through by Ψ^{-1} and using the correct normalisation)

$$q^{\frac{1}{2}}\Psi - q^{-\frac{1}{2}}\Psi^{-1} = (q - q^{-1})\mathrm{id}.$$

The incorrect factors in front of Ψ, Ψ^{-1} are related to the fact that we
still need to correct this by the writhe to obtain an invariant of ambient
isotopy. But apart from this, it is clear that we obtain a version of the
Jones invariant.

Finally, some general remarks. First of all, the multiplicative quantum
dimension $\underline{\dim}'(V)$ studied in the last lecture is the invariant for framed
links associated to V, evaluated on $\bigcirc$. We can similarly have the
'trefoil dimension' by evaluating on the trefoil knot etc.

Secondly, since every finite-dimensional ribbon quasitriangular Hopf
algebra H is an object in its own category of modules by the left regular
representation, we have a natural invariant of framed links associated to
any one of these. In fact, we have a kind of 'pairing' between links and
such Hopf algebras in this way. Similarly, the quantum order is

$$|H| = \langle H, \, \text{⌇} \rangle$$

from the earlier point of view, as a pairing between quasitriangular Hopf
algebras and regular isotopy classes. I.e. we can interpret such expressions equally well either as an invariant of links defined by a suitable

Hopf algebra or as an invariant of such Hopf algebras defined by a suitable link. For example, one can have the 'trefoil order' associated to the trefoil knot.

14

Hopf algebras in braided categories.
Coaddition on $\mathbb{A}_q^2$

Whereas the last lecture was about using Hopf algebras to do knot theory, this lecture is about the other side of the pairing, using knots to do algebra. This is going to be another 'fun' lecture but actually the applications of these ideas are quite serious and lead to

  – inhomogeneous quantum groups, such as parabolic ones
  – quantum groups $U_q(\mathfrak{g})$ for general complex semisimple $\mathfrak{g}$

and many other constructions. Note that I could just write down a whole bunch of generators and relations for $U_q(\mathfrak{g})$, but I'd like instead to give some insight into what is really going on.

First of all, we have seen so far 'linear algebra' in braided categories. Let us now start to do 'algebra'. Clearly,

Definition 14.1 *In any monoidal category $\mathcal{C}$, an algebra B is*

1. An object B of $\mathcal{C}$.

2. A product morphism $m : B \otimes B \to B$ obeying an associativity condition. In an obvious extension $m = \curlyvee$ of our diagrammatic notation, this means the equality

3. A unit morphism $\eta : \underline{1} \to B$ obeying the usual unity axioms but now in the category $\mathcal{C}$. Writing η as a morphism from nothing (which is how we write $\underline{1}$) to B, we require

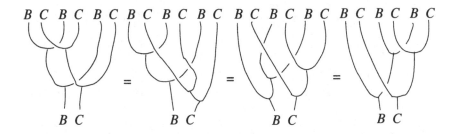

Fig. 14.1. Tensor product of two braided algebras.

For example, an H-module algebra is nothing other than an algebra in $_H\mathcal{M}$, where H is a Hopf algebra. To go further, one needs to assume a braiding to play the role of any hidden transpositions in algebraic constructions.

Lemma 14.2 *Let B, C be two algebras in a braided monoidal category. Then the object $B \otimes C$ also has the structure of an algebra in the category, denoted by $B\underline{\otimes}C$, the braided tensor product algebra, and defined by*

$$m_{B\underline{\otimes}C} = (m_B \otimes m_C) \circ (\mathrm{id} \otimes \Psi_{C,B} \otimes \mathrm{id})$$

and the tensor product unit morphism.

Proof That the product is associative is shown in Figure 14.1 using the diagrammatic notation. The left hand side of the figure is the product of $B\underline{\otimes}C$ used twice in one order, and the right hand side is the product used twice in the other order. The first equality is functoriality under the product morphism $C \otimes C \to C$, which allows us to push that down over the rightmost copy of B. Then we use associativity in B and C to reorganise the branches. After this, we use functoriality again to push the product morphism of B up and under the leftmost copy of C. The proof of the unit is more trivial and is left as an exercise. $\square$

The reader can have some fun checking that for three algebras B, C, D in the category, one has $(B\underline{\otimes}C)\underline{\otimes}D \cong B\underline{\otimes}(C\underline{\otimes}D)$ via the underlying associativity Φ (which we are suppressing). Of course, the construction models the usual tensor product of algebras or supertensor product of superalgebras if one knows about those.

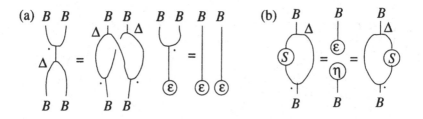

Fig. 14.2. Main axioms of a braided group or Hopf algebra in a braided category.

Definition 14.3 *A braided group* $(B, m, \eta, \Delta, \epsilon, S)$ *or Hopf algebra in a braided category is*

1. An algebra (B, m, η) *in the category.*

2. Morphisms $\Delta : B \to B \underline{\otimes} B$, $\epsilon : B \to \underline{1}$ *forming a coalgebra (axioms as for an algebra but diagrams turned upside-down).*

3. Δ, ϵ *are algebra maps, where* $B \underline{\otimes} B$ *has the braided tensor product algebra structure (see Figure 14.2(a)).*

4. A morphism $S : B \to B$ *obeying the usual axioms but now as morphisms in the braided category (see Figure 14.2(b)).*

One also considers bialgebras in braided categories, i.e. without the antipode S. In fact, it turns out that all the elementary Hopf algebra theory from our early lectures can be developed in this diagrammatic setting. We will just give a small taste of this.

Lemma 14.4 *The antipode of any braided group* B *obeys*

$$S \circ \cdot = \cdot \circ \Psi_{B,B} \circ (S \otimes S), \quad \Delta \circ S = (S \otimes S) \circ \Psi_{B,B} \circ \Delta.$$

Proof This is shown in Figure 14.3. We use the unit and counit axioms to graft on two loops involving S, knowing that they collapse to $\eta \circ \epsilon$ from Figure 14.2(b). After some reorganisation, we use the homomorphism property of Δ in Figure 14.2(a) and then Figure 14.2(b) again for the final result. For the second part of the lemma, just turn the page upside down and read the diagrammatic proof again! □

The most beautiful thing about braided groups is that they lie right on the interface between algebra and knot theory, i.e. hopefully it is agreed that these proofs are 'fun'. On the other hand, we can do most things in group theory in this setting. One can keep in mind the group

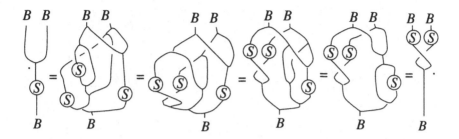

Fig. 14.3. Proof of the braided antihomomorphism property of S.

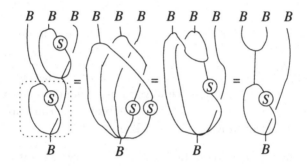

Fig. 14.4. Proof of the braided adjoint action (in box) of a braided group B on itself.

Hopf algebra kG viewed as a braided group with the trivial braiding and coproduct $\Delta g = g \otimes g$ for $g \in G$, etc. Then we can just write out our favourite group theory construction as a sequence of morphisms, including any hidden transpositions. The construction then usually (but not always) generalises to the braided case. For example:

Proposition 14.5 *For any braided group B, there is an* adjoint action $\mathrm{Ad} : B \otimes B \to B$ *of B on itself, shown in the box in Figure 14.4.*

Proof The left side of Figure 14.4 is $\mathrm{Ad} \circ (\mathrm{id} \otimes \mathrm{Ad})$ while the right hand side is the product of B and then the application of Ad. Equality of these is clearly what we mean by an action. We use functoriality to pull down some of the products to the bottom. The analogue in our notation

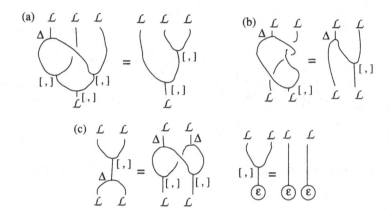

Fig. 14.5. A braided-Lie algebra, as extracted from Ad.

of 'dropping brackets' or 'renumbering to linear numbering' for Hopf algebras is to identify the possible ways to make multiple products as a multiple branch, as shown. We then use the braided antimultiplicativity property of the braided antipode from Lemma 14.4.   □

This concludes our taste of braided group theory. This is not really what the lecture is about so let us just mention one aside that we will come back to at the end of the course. Namely, by elaborating the various properties of Ad along the same lines as above, one can discard the braided group itself and define a *braided-Lie* algebra as $(\mathcal{L}, \Delta, \epsilon, [\ ,\ ])$ where

1. $(\mathcal{L}, \Delta, \epsilon)$ is a coalgebra in the category.
2. $[\ ,\ ] : \mathcal{L} \otimes \mathcal{L} \to \mathcal{L}$ obeys axioms extracted from the properties of Ad, as shown in Figure 14.5.

One can show that associated to a braided-Lie algebra in a category with suitable direct sums etc., one has a braided group $U(\mathcal{L})$ (without antipode). This is a certain quotient of the tensor algebra over $\mathcal{L}$ and has coproduct given by Δ extended to products. It is beyond our scope, but there is for example a braided-Lie algebra $\widetilde{sl}_{2,q}$ whose enveloping braided group is a homogenised and braided version of $U_q(sl_2)$. Also, for a usual Lie algebra $\mathfrak{g}$ one can check that

$$\mathcal{L} = k1 \oplus \mathfrak{g}, \quad \Delta 1 = 1 \otimes 1, \quad \Delta x = x \otimes 1 + 1 \otimes x, \quad [1, x] = x, \quad [x, 1] = 0$$

and the initial Lie bracket for $x, y \in \mathfrak{g}$ is a braided-Lie algebra with Ψ

trivial. In this case $U(\mathcal{L})$ is a homogenised version of $U(\mathfrak{g})$. Similarly, if one wants a natural formulation of a finite-dimensional 'Lie object' underlying $U_q(\mathfrak{g})$ then this is it.

We now come down to earth a bit with some simple examples. Recall from Example 11.2 that $k_q\mathbb{Z}$ (the group algebra of $\mathbb{Z}$ equipped with a dual quasitriangular structure) generates as its comodule category that of $\mathbb{Z}$-graded spaces, with Ψ given by q and the degrees. Here $q \in k^*$.

Example 14.6 $\mathbb{A}_q^1 = k[x]$ *with* $|x| = 1$ *and*

$$\Delta x = x \otimes 1 + 1 \otimes x, \quad \epsilon(x) = 0, \quad Sx = -x$$

is a braided group (the braided line*) in the category of $\mathbb{Z}$-graded spaces. Here*

$$|x^n| = n, \quad \Psi(x^m \otimes x^n) = q^{mn} x^n \otimes x^m.$$

It is a nice exercise to show that

$$\Delta x^n = \sum_{r=0}^{n} \begin{bmatrix} n \\ r \end{bmatrix}_q x^r \otimes x^{n-r}, \quad Sx^n = q^{\frac{n(n-1)}{2}}(-1)^n x^n.$$

Proof Elementary. One has to check that Δ extends as a homomorphism $B \to B \underline{\otimes} B$. However, this is $k[x] \otimes k[y]$ with relations $yx = qxy$ due to the form of Ψ. Thus,

$$(1 \otimes x)(x \otimes 1) = \Psi(x \otimes x) = qx \otimes x = q(x \otimes 1)(1 \otimes x)$$

and we identify $x \equiv x \otimes 1$ and $y \equiv 1 \otimes x$. We then use the q-binomial theorem for the form of Δ on products. As for the form of S on products, we use Lemma 14.4 and the form of the braiding. We take this as a definition of S on products. We then verify the axioms. $\square$

Similarly, we have:

Example 14.7 *The quantum or 'braided' plane $\mathbb{A}_q^2$ with relations $yx = qxy$ is a braided group with*

$$\Delta x = x \otimes 1 + 1 \otimes x, \quad \Delta y = y \otimes 1 + 1 \otimes y,$$

$$\epsilon x = \epsilon y = 0, \quad Sx = -x, \quad Sy = -y,$$

$$\Psi(x \otimes x) = q^2 x \otimes x, \quad \Psi(x \otimes y) = qy \otimes x, \quad \Psi(y \otimes y) = q^2 y \otimes y,$$

$$\Psi(y \otimes x) = qx \otimes y + (q^2 - 1)y \otimes x.$$

Proof The proof by direct verification is similar to the above. One has to check that Δ extends as a homomorphism to the braided tensor product, etc. We leave this as an exercise. Note that the braiding here is proportional to the one induced in $\mathbb{A}_q^2$ as a comodule algebra under $SL_q(2)$. We have already seen that the latter is dual quasitriangular, hence its category of comodules is braided. $\square$

15

Braided differentiation

We have seen in the last lecture that one really needs this diagrammatic theory of braided groups (as opposed to quantum groups) to describe even such simple things as the additive 'coaddition' structure on the quantum plane. From what was given in the last lecture it is a nice exercise to find that it has the explicit form

$$\Delta(x^m y^n) = \sum_{r=0}^{m} \sum_{s=0}^{n} \begin{bmatrix} m \\ r \end{bmatrix}_{q^2} \begin{bmatrix} n \\ s \end{bmatrix}_{q^2} x^r y^s \otimes x^{m-r} y^{n-s} q^{(m-r)s}$$

on a general basis element of $\mathbb{A}_q^2$.

Now, what can one do with a coaddition? Well, in usual geometry or algebraic geometry one can make an infinitesimal addition and define differentiation in this way. The general idea is the following braided group version of the coregular action R^* which we have seen in Lecture 2.

First of all, if B is rigid, we have duals of all its structure morphisms (as defined in Lecture 12), which clearly define a dual braided group B^*. This is shown in Figure 15.1.

More generally, we say that two braided groups B', B are dually paired if there is a morphism ev : $B' \otimes B \to \underline{1}$ (we still write ev = $\smile$) such that the product of one corresponds to the coproduct of the other. The diagrams are obtained by dragging the lower legs of the definitions of Δ^*, m^* etc. out and up to the right. See Figure 15.2(a). Note that this categorical definition does not have an unnecessary and unnatural extra braiding which would be present if we just mimicked the pairing of ordinary Hopf algebras. One still has a coregular action by evaluation against the coproduct, which we describe next.

Lemma 15.1 *If B is a braided group with invertible antipode in a braided*

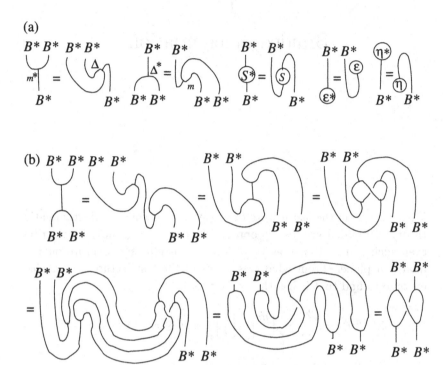

Fig. 15.1. Dual braided group.

category C *then* B^{cop} *consisting of the same algebra as* B *but the coproduct* $\Psi^{-1} \circ \Delta_B$ *is a braided group in the category* $\bar{C}$ *with reversed braiding.*

Proof Elementary diagrammatics. A braided group in $\bar{C}$ means, in terms of our diagrammatic notations for C, that B obeys the axioms of a braided group as given in the last lecture, but with $\times$ and $\times$ interchanged. $\qquad\square$

Proposition 15.2 *If* B' *is dually paired with* B *then* B *becomes a braided* B'^{cop}-*module algebra in the category* $\bar{C}$ *with reversed braiding, by evaluating* B' *against the coproduct of* B *(the left coregular action).*

Proof This is more fun with diagrams and shown in Figure 15.2(b).

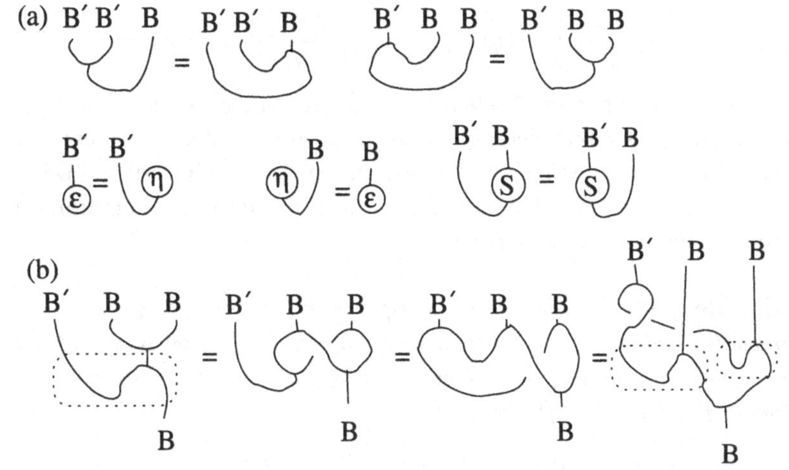

Fig. 15.2. Coregular action on a braided group by one dually paired with it.

The box on the left is the (left) coregular action. It is easy to see that it is indeed a left action. What is on the right is two applications of the coaction. We see also the coproduct of B'^{cop}. With respect to this, this is what one means by a braided module algebra (i.e. the product commutes with the action, in this case of B'^{cop}).   □

What this boils down to in our simple examples of the braided line and the quantum-braided plane is the following. On the braided line $B = \mathbb{A}_q^1 = k[x]$ (in the category of $k_q\mathbb{Z}$-comodules) define the operation $\partial_q : B \to B$ by

$$\partial_q f = \text{coeff}_{x \otimes}(\Delta f), \quad \text{i.e.} \quad \Delta f = 1 \otimes f + x \otimes \partial_q f + \cdots$$

where $\cdots$ denotes higher powers of x. There is a suggestive way to write this, and compute it. Namely, recall that Δ is a homomorphism $B \to B \underline{\otimes} B$. In our case the latter is $k[x] \underline{\otimes} k[y] = \mathbb{A}_q^2$ the quantum plane. To see this, write $x = x \otimes 1$ and $y = 1 \otimes x$ (we denote the generator of the second copy by y). Hence

$$yx = (1 \otimes x)(x \otimes 1) = \Psi(x \otimes x) = qx \otimes x = q(x \otimes 1)(1 \otimes x) = qxy$$

by the definition of the braided tensor product. In this notation one can easily obtain Δx^n as given in the last lecture, via the q-binomial formula. Then

$$\partial_q f(y) = x^{-1}(f(x+y) - f(y))|_{x=0} = \frac{f(y) - f(qy)}{(1-q)y}; \quad \partial_q x^m = [m]_q x^{m-1}$$

and the usual differential when $q = 1$. Of course, we are not really dividing by x here but picking out the term in the difference linear in x. This is the correct setting for the q-derivative, as an infinitesimal braided addition or 'braided differentiation'. It also fits into our diagrammatic theory of the coregular representation.

Example 15.3 *The braided line* $\mathbb{A}^1_q = k[x]$ *in the category of* $k_q\mathbb{Z}$-*comodules is dually paired with another copy* $k[y]$ *(say) by* $\mathrm{ev}(y \otimes x) = 1$, *where* $|y| = -1$, *and* ∂_q *is the action of* y *in the coregular representation. Moreover, Figure 15.2(b) becomes*

$$\partial_q(fg) = (\partial_q f)g + m \circ \Psi^{-1}(\partial_q \otimes f)g = (\partial_q f)g + f(qx)\partial_q g,$$

where $\Psi^{-1}(\partial_q \otimes x^m) = q^m x^m \otimes \partial_q$ *as in the braiding with* y *(more precisely, we should write* $\Psi^{-1}(y \otimes x^m) = q^m x^m \otimes y$ *followed by the action of* y *as* ∂_q*).*

Proof Elementary from the form of the coproduct and the braiding of these braided groups. The degrees $|x| = 1$, $|y| = -1$ (i.e. the coactions $\beta(x) = x \otimes g$, $\beta(y) = y \otimes g^{-1}$ determine the braiding $\Psi(x \otimes y) = q^{-1}y \otimes x$, the inverse of which we use here. One has to check that the pairing shown extends as a pairing of braided groups. $\square$

Whereas this example is trivial enough to discover 'by hand', the same ideas apply for much more complicated braided groups, such as the quantum plane and its higher-dimensional analogues. First of all:

Example 15.4 *The partial derivatives* $\partial_{q,x}, \partial_{q,y} : \mathbb{A}^2_q \to \mathbb{A}^2_q$ *are the operators defined by*

$$\Delta f = 1 \otimes f + x \otimes \partial_{q,x} f + y \otimes \partial_{q,y} f + \cdots$$

where $\cdots$ *denotes terms of higher total degree in the left* $\otimes$ *factor. Explicitly,*

$$\partial_{q,x} x^m y^n = [m]_{q^2} x^{m-1} y^n, \quad \partial_{q,y} x^m y^n = q^m [n]_{q^2} x^m y^{n-1}.$$

Moreover, $\partial_{q,y}\partial_{q,x} = q^{-1}\partial_{q,x}\partial_{q,y}.$

Proof This follows from the form of the additive coproduct of $\mathbb{A}^2_q$. $\square$

Then, underlying this is the braided group coregular representation again. Here $B' = k\langle v, w \rangle$ modulo the relations $wv = q^{-1}vw$ is a quantum plane $\mathbb{A}_{q^{-1}}^2$, but with the coaction

$$\beta(v^i) = \sum_a v^a \otimes St^i{}_a,$$

where

$$t^i{}_j = \begin{pmatrix} a & b \\ c & d \end{pmatrix}, \quad v^i = \begin{pmatrix} v \\ w \end{pmatrix}$$

are the matrix generators of $SL_q(2)$ and $\mathbb{A}_{q^{-1}}^2$. This is the quantum plane as a 'vector' right comodule algebra (the coaction is the dual coaction to the coaction on (x, y) as a 'covector' right comodule algebra under $SL_q(2)$). It forms a braided group with the coproduct and braiding

$$\Delta v = v \otimes 1 + 1 \otimes v, \quad \Delta w = w \otimes 1 + 1 \otimes w$$

$$\epsilon(v) = \epsilon(w) = 0, \quad Sv = -v, \quad Sw = -w$$

$$\Psi(v \otimes v) = q^2 v \otimes v, \quad \Psi(v \otimes w) = qw \otimes v + (q^2 - 1)v \otimes w$$

$$\Psi(w \otimes v) = qv \otimes w, \quad \Psi(w \otimes w) = q^2 w \otimes w$$

by a similar computation as for $\mathbb{A}_q^2$.

As for $\mathbb{A}_q^2$, the braiding here is only proportional to the one as an object in the category of $SL_q(2)$-comodules. To take care of the incorrect normalisation, we define

$$\widetilde{SL_q(2)} = SL_q(2) \otimes k_{\lambda^{-1}}\mathbb{Z}, \quad \lambda = q^{-\frac{3}{2}}$$

as a tensor product of dual quasitriangular Hopf algebras. Then the coactions are

$$\beta(x_i) = \sum_a x_a \otimes t^a{}_i g, \quad \beta(v^i) = \sum_a v^a \otimes St^i{}_a g^{-1}$$

where $k_{\lambda^{-1}}\mathbb{Z} = k[g, g^{-1}]$. These allow us the consider $\mathbb{A}_q^2$ and $\mathbb{A}_{q^{-1}}^2$ correctly in the braided category of $\widetilde{SL_q(2)}$-comodules and induce the stated braidings.

Example 15.5 $B' = \mathbb{A}_{q^{-1}}^2$ *is dually paired with* $B = \mathbb{A}_q^2$ *by*

$$\mathrm{ev}(v \otimes x) = 1, \quad \mathrm{ev}(v, y) = 0, \quad \mathrm{ev}(w, x) = 0, \quad \mathrm{ev}(w, y) = 1$$

as braided groups in the category of right $\widetilde{SL_q(2)}$-comodules, and $\partial_{q,i}$ are the action of v^i in the coregular representation. Hence

$$\partial_{q,i}(fg) = (\partial_{q,i}f)g + m \circ \Psi^{-1}(\partial_{q,i} \otimes f)g,$$

where $\Psi^{-1}(\partial_{q,i} \otimes f)$ stands for $\Psi^{-1}(v^i \otimes f)$ followed by the action of the resulting v^j as $\partial_{q,j}$.

Proof We clearly have two braided groups; one has to check that they are dually paired by ev as stated, extended to products as a pairing of braided groups. This can again be done by explicit computation – but there are general R-matrix methods which we will come to later on. Once the categorical picture is established, we obtain the braiding between $\mathbb{A}_q^2$ and $\mathbb{A}_{q^{-1}}^2$ from the stated coactions and the dual quasitriangular structure of $\widetilde{SL_q(2)}$.   □

What clearly emerges here is an entire 'geometry' taking place in the braided category of $SL_q(2)$-comodules (or $\widetilde{SL_q(2)}$-comodules for the entire braided group structure). What about matrices in this category? Unfortunately, the coaction Ad_R of a quantum group on itself, while respecting the coalgebra structure, does not usually respect the algebra structure as soon as the algebra is noncommutative.

Example 15.6 *Let $q \in k^*$. There is a covariant version $B = BSL_q(2)$ of $SL_q(2)$ defined as $k\langle \alpha, \beta, \gamma, \delta \rangle$ modulo 'braided commutativity' relations*

$$\beta\alpha = q^2\alpha\beta, \quad \gamma\alpha = q^{-2}\alpha\gamma, \quad \delta\alpha = \alpha\delta, \quad \beta\gamma = \gamma\beta + (1 - q^{-2})\alpha(\delta - \alpha),$$

$$\delta\beta = \beta\delta + (1 - q^{-2})\alpha\beta, \qquad \gamma\delta = \delta\gamma + (1 - q^{-2})\gamma\alpha,$$

and the 'braided determinant' relation $\alpha\delta - q^2\gamma\beta = 1$. The coaction of $SL_q(2)$ is

$$\beta(u^i{}_j) = \sum_{a,b} u^a{}_b \otimes (St^i{}_a)t^b{}_j; \quad u^i{}_j = \begin{pmatrix} \alpha & \beta \\ \gamma & \delta \end{pmatrix}.$$

Moreover, $BSL_q(2)$ forms a braided group with

$$\Delta u^i{}_j = \sum_a u^i{}_a \otimes u^a{}_j, \quad \epsilon(u^i{}_j) = \delta^i{}_j$$

and the braiding in the category induced by this coaction.

Proof This comes out of a general construction called transmutation, which we will outline later. For the moment it is just an example, which may be verified explicitly. Again, there are general R-matrix methods for this kind of algebra. $\square$

There is also an antipode which we do not give explicitly. (Without the braided-determinant relation, this is the braided matrices $BM_q(2)$.) There is also a braided version of $GL_q(2)$. In a matrix multiplication notation, one can write $\beta(u) = t^{-1}ut$ as an element of $BSL_q(2) \otimes SL_q(2)$.

We have put in this example (leaving out the rather tedious direct verifications) to stress the idea that in the braided category of $\widetilde{SL_q(2)}$ we have a whole 'universe' of objects. The general principle is that whatever category is chosen determines versions of lines, planes, differentials, matrices etc., i.e. a whole 'braided geometry'. I.e. it is far from unique.

Let us end with some general remarks. We know that every quasitriangular Hopf algebra generates a braided category of modules. Moreover, a module algebra B under H just means an algebra in this braided category. We can therefore obtain many results about module algebras by our braid diagram methods.

Corollary 15.7 *Let B, C be two module algebras under a quasitriangular Hopf algebra H. There is a braided tensor product $B \underline{\otimes} C$ as an H-module algebra. It is built on $B \otimes C$ with product*

$$(a \otimes c)(b \otimes d) = \sum a\mathcal{R}^{(2)} \triangleright b \otimes (\mathcal{R}^{(1)} \triangleright c)d, \quad \forall a, b \in B, \quad c, d \in C.$$

Proof This is an immediate corollary of the construction proven in the last lecture, in Figure 14.1. $\square$

Similarly for dual quasitriangular Hopf algebras and comodule algebras under them. In the next lecture we will continue this idea of using braid diagrams to obtain results about ordinary Hopf algebras (the bosonisation theorem). Moreover, what should be clear by now is that the 'geometry' with (dual) quasitriangular quantum groups as symmetries is intrinsically braided: anything on which such a quantum group (co)acts covariantly is braided, i.e. lives in a braided category.

16

Bosonisation. Inhomogeneous quantum groups

We have completed our introduction to 'braided geometry' with the basic constructions and basic examples. We are now ready to apply these methods back to ordinary Hopf algebras. The construction in this lecture is called 'bosonisation' because it turns a braided group (which is a generalisation of a supergroup) into an ordinary Hopf algebra (such as an ordinary group) [the term comes from physics].

First of all, a general concept which we will use. It is just for clarity (we are not going to do any heavy category theory). Thus, a monoidal functor $F : \mathcal{C} \to \mathcal{V}$ between monoidal categories is a pair (F, c) where

1. F is a functor.
2. $c : F^2 \to F \circ \otimes$ is a natural isomorphism between the functors $F^2, F \circ \otimes : \mathcal{C} \times \mathcal{C} \to \mathcal{V}$ (here $F^2(V, W) = F(V) \otimes F(W)$). This is a collection of functorial isomorphisms

$$c_{V,W} : F(V) \otimes F(W) \cong F(V \otimes W).$$

3. The condition in Figure 16.1 holds.
4. We also require

$$F(\underline{1}) = \underline{1}, \quad c_{\underline{1},V} \circ l_{F(V)} = F(l_V), \quad c_{V,\underline{1}} \circ r_{F(V)} = F(r_V)$$

for compatibility with the unit object.

This is more or less obvious, and in fact in our present applications, Φ will be the trivial vector space associativity and c will also be the trivial vector space identification. So we are just saying that F respects $\otimes$.

Lemma 16.1 *Let H be a quasitriangular Hopf algebra. There is a monoidal functor*

$$_H\mathcal{M} \hookrightarrow {}^H_H\mathcal{M}, \quad (V, \triangleright) \mapsto (V, \triangleright, \beta), \quad \beta(v) = \sum \mathcal{R}^{(2)} \otimes \mathcal{R}^{(1)} \triangleright v$$

$$(F(V) \otimes F(W)) \otimes F(Z) \xrightarrow{\ c \otimes \mathrm{id}\ } F(V \otimes W) \otimes F(Z) \xrightarrow{\ c\ } F((V \otimes W) \otimes Z)$$

$$\Big\downarrow \Phi \qquad\qquad\qquad\qquad\qquad\qquad\qquad\qquad \Big\downarrow F(\Phi)$$

$$F(V) \otimes (F(W) \otimes F(Z)) \xrightarrow{\ \mathrm{id} \otimes c\ } F(V) \otimes F(W \otimes Z) \xrightarrow{\ c\ } F(V \otimes (W \otimes Z))$$

Fig. 16.1. A monoidal functor F.

for all $v \in V$. Here β is said to be the coaction 'induced' by an action on any module V.

Proof This is a nice exercise from the axioms of a quasitriangular structure. That β is indeed a coaction is the $(\mathrm{id} \otimes \Delta)\mathcal{R}$ axiom. That it forms a crossed module with the original action is

$$h_{(1)} v^{(\bar{1})} \otimes h_{(2)} \triangleright v^{(\bar{2})} = h_{(1)} \mathcal{R}^{(2)} \otimes (h_{(2)} \mathcal{R}^{(1)}) \triangleright v$$
$$= \mathcal{R}^{(2)} h_{(2)} \otimes (\mathcal{R}^{(1)} h_{(1)}) \triangleright v = (h_{(1)} \triangleright v)^{(\bar{1})} h_{(2)} \otimes (h_{(1)} \triangleright v)^{(\bar{2})}$$

for all $v \in V$ and $h \in H$, using the quasicocommutativity axiom of a quasitriangular structure. That the functor is monoidal, i.e. respects the tensor products on both sides, is the remaining $(\Delta \otimes \mathrm{id})\mathcal{R}$ axiom, since $\beta(v \otimes w) = \mathcal{R}^{(2)} \otimes \mathcal{R}^{(1)} \triangleright (v \otimes w) = \mathcal{R}^{(2)} \otimes \mathcal{R}^{(1)}{}_{(1)} \triangleright v \otimes \mathcal{R}^{(1)}{}_{(2)} \triangleright w$. $\qquad\square$

We also need some obvious constructions which we could have as well done in the first lectures.

Lemma 16.2 *Let H be a Hopf algebra or bialgebra and B a left H-module algebra. Then there is a cross product (also called 'smash product') algebra $B \rtimes H$ defined by*

1. $B \otimes H$ as a vector space.

2. The product

$$(b \otimes h)(c \otimes g) = \sum b(h_{(1)} \triangleright c) \otimes h_{(2)} g, \quad \forall b, c \in B, \ h, g \in H.$$

3. The tensor product unit.

Proof Elementary exercise to verify associativity. $\qquad\square$

Reversing all arrows we have equally well:

Lemma 16.3 *Let H be a Hopf algebra or bialgebra and C a left H-*

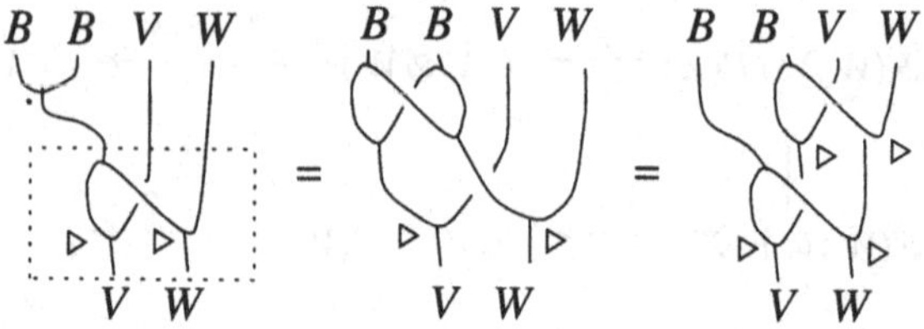

Fig. 16.2. Tensor product of braided modules.

comodule coalgebra. Then there is a cross coproduct *algebra* $B \bowtie H$
defined by

1. $B \otimes H$ *as a vector space.*

2. *The coproduct*

$$\Delta(b \otimes h) = \sum b_{(1)} \otimes {b_{(2)}}^{(\bar{1})} \otimes h_{(1)} \otimes {b_{(2)}}^{(\bar{2})} \otimes h_{(2)}, \quad \forall b \in B, \ h \in H.$$

3. $\epsilon(b \otimes h) = \epsilon(b)\epsilon(h), \quad \forall b \in B, \ h \in H.$

Proof Also an exercise.   □

Finally, we need the concept of a braided B-module where B is a
braided group in a braided category $\mathcal{C}$. Clearly this means an object V
in the category and a morphism $B \otimes V \to V$ obeying the obvious conditions. The category $_B\mathcal{C}$ of braided B-modules in $\mathcal{C}$ is itself monoidal, i.e.
given two such modules V, W we have clearly a tensor product braided
B-module $V \otimes W$ as shown in Figure 16.2.

Theorem 16.4 *(Bosonisation theorem) Let H be a quasitriangular Hopf
algebra and $B \in {_H\mathcal{M}}$ a braided group. Then there is an associated
ordinary Hopf algebra $B \bowtie H$, the* bosonisation *of B, defined by*

1. $B \bowtie H$ *as an algebra by the given action of H on B as an object.*

2. $B \bowtie H$ *as a coalgebra by the induced coaction in Lemma 16.1.*

*Moreover, the braided modules of B are in 1-1 correspondence with the
ordinary modules of $B \bowtie H$ (an identification of monoidal categories).*

Proof We already have algebras and coalgebras by Lemmas 16.2 and 16.3.
Using the explicit description in the lemmas, we check that the two fit

together to form a bialgebra. Thus,

$$\Delta((1 \otimes h)(b \otimes 1)) = (h_{(1)} \triangleright b)_{(1)} \otimes \mathcal{R}^{(2)} h_{(2)} \otimes \mathcal{R}^{(1)} \triangleright (h_{(1)} \triangleright b)_{(2)} \otimes h_{(3)}$$
$$= h_{(1)} \triangleright b_{(1)} \otimes \mathcal{R}^{(2)} h_{(3)} \otimes (\mathcal{R}^{(1)} h_{(2)}) \triangleright b_{(2)} \otimes h_{(4)}$$

since the coproduct $B \to B \otimes B$ is a morphism, i.e. commutes with the action of H. On the other hand,

$$(\Delta(1 \otimes h))(\Delta(b \otimes 1)) = h_{(1)} \triangleright b_{(1)} \otimes h_{(2)} \mathcal{R}^{(2)} \otimes (h_{(3)} \mathcal{R}^{(1)}) \triangleright b_{(2)} \otimes h_{(4)}$$

on using the definition of the coproduct and product of $B {\rtimes} H$. The two are equal by the quasicocommutativity axiom.

Moreover, given a braided B-module V, we have an action of B on V and also an action of H on V since by definition this is an object in our braided category $_H\mathcal{M}$. The corresponding action of $B {\rtimes} H$ is

$$(b \otimes h) \triangleright v = b \triangleright (h \triangleright v), \quad \forall v \in V, \ b \in B, \ h \in H.$$

We require

$$h \triangleright (b \triangleright v) = (1 \otimes h)(b \otimes 1) \triangleright v = (h_{(1)} \triangleright b \otimes h_{(2)}) \triangleright v = (h_{(1)} \triangleright b) \triangleright (h_{(2)} \triangleright v)$$

which is precisely the condition that $\triangleright : B \otimes V \to V$ is a morphism in $_H\mathcal{M}$. Conversely, given an action of $B {\rtimes} H$ we pull back to the two subalgebras B, H and obtain a braided B-module.

Morphisms between braided B-modules are also H-module maps being morphisms in $_H\mathcal{M}$, and clearly correspond to $B {\rtimes} H$-module maps, i.e. we have a functorial identification. Finally, this functor is monoidal. Thus, given braided modules V, W, the action of B from Figure 16.2 is

$$b \otimes v \otimes w \mapsto b_{(1)} \otimes \mathcal{R}^{(2)} \triangleright v \otimes \mathcal{R}^{(1)} \triangleright b_{(2)} \otimes w \mapsto b_{(1)} \triangleright (\mathcal{R}^{(2)} \triangleright v) \otimes (\mathcal{R}^{(1)} \triangleright b_{(2)}) \triangleright w$$

on elements $v \in V$, $w \in W$ and $b \in B$ (reading down the figure and using the braiding of $_H\mathcal{M}$). Under the above correspondence, this is

$$(b_{(1)} \otimes \mathcal{R}^{(2)}) \triangleright v \otimes (\mathcal{R}^{(1)} \triangleright b_{(2)} \otimes 1) \triangleright w = (b \otimes 1) \triangleright (v \otimes w)$$

when we compare with the coproduct of $B {\rtimes} H$. The other parts of the proof are more trivial and we omit them. $\qquad \square$

The cross relations and coproduct in the theorem can be written more concisely as

$$hb = \sum (h_{(1)} \triangleright b) h_{(2)}, \quad \Delta b = \sum b_{(1)} \mathcal{R}^{(2)} \otimes \mathcal{R}^{(1)} \triangleright b_{(2)}$$

for all $h \in H$, $b \in B$ when viewed in $B {\rtimes} H$ by the identifications $B = B \otimes 1$ and $H = 1 \otimes H$. We can also give an immediate example, as follows. We have already seen the braided line $\mathbb{A}_q^1$ which is $k[x]$ viewed as a braided group in $k_q\mathbb{Z}$. It has a natural quotient when q is a primitive n'th root of unity, namely

$$B = k[x]/\langle x^n \rangle, \quad \Delta x = x \otimes 1 + 1 \otimes x, \quad \epsilon x = 0, \quad Sx = -x.$$

It lives in the category of $k_q\mathbb{Z}/n$-modules by the action $g \triangleright x = qx$. The case $n = 2$ is called a 'superline' [or in physics 'Grassmann variable'] while the general case is called the 'anyonic line'. It is a reduced version of the braided line $\mathbb{A}_q^1$ in Example 14.6.

Example 16.5 *Let q be a primitive n'th root of 1. The bosonisation of $B = k[x]/\langle x^n \rangle$ is a version of $u_q(b_+)$, the Borel sub-Hopf algebra of $u_q(sl_2)$.*

Proof A nice exercise from the definitions. One obtains $B \rtimes k_q\mathbb{Z}/n = k\langle x, g \rangle$ modulo the relations $gx = qxg$ and with the coproduct $\Delta x = x \otimes 1 + g \otimes x$ as in Lecture 1. $\square$

There is equally well a right-module version of the bosonisation theorem which gives exactly $u_q(b_+)$ in the conventions of Lecture 6 (on replacing q^2 there by q). Similarly, let $B = k[F]/\langle F^n \rangle$ be the braided group in the category of left $k_q\mathbb{Z}/n$-modules by $g \triangleright F = q^{-1}F$. In this case the bosonisation theorem above gives

$$gF = q^{-1}Fg, \quad \Delta F = F \otimes 1 + g^{-1} \otimes F, \quad \epsilon F = 0,$$

which is basically $u_q(b_-)$, the other Borel sub-Hopf algebra. In the next lecture we will see how these are naturally glued together to obtain $u_q(sl_2)$ itself.

We can also reverse all arrows of course. Without redoing the proofs, suffice it to say that if H is a dual quasitriangular Hopf algebra, there is a monoidal functor

$$\mathcal{M}^H \hookrightarrow \mathcal{M}_H^H, \quad (V, \beta) \mapsto (V, \beta, \triangleleft), \quad v \triangleleft h = v^{(1)} \mathcal{R}(v^{(2)} \otimes h)$$

for all $v \in V$, $h \in H$. We have switched to right modules and comodules here – there is an equally good version with left modules and comodules. If $B \in \mathcal{M}^H$ is a braided group, then there is an ordinary Hopf algebra $H \ltimes B$ by the given coaction and the induced action via this functor. As an example, the bosonisation of the braided line $\mathbb{A}_q^1$ in the category of right $k_q\mathbb{Z}$-comodules is a version of the Hopf algebra $U_q(b_+)$. Since we use the dual version here and yet still obtain the same kind of Hopf algebra, we see the reason for the self-duality pairing of $U_q(b_+)$ with itself which we have seen back in Lecture 2: it is a reflection under bosonisation of the self-duality pairing of the braided line with itself!

Finally, we can use bosonisation to construct many new and interesting quantum groups.

Example 16.6 *The bosonisation of the quantum plane is the inhomogeneous quantum group $\widetilde{SL_q(2)} {\ltimes} \mathbb{A}_q^2$. Writing a, b, c, d, g, g^{-1} for the generators of $\widetilde{SL_q(2)}$, and X, Y for those of $\mathbb{A}_q^2$, the cross relations and coproduct are*

$$Xg = q^{\frac{3}{2}}gX, \quad Yg = q^{\frac{3}{2}}gY, \quad X\begin{pmatrix} a & b \\ c & d \end{pmatrix} = \begin{pmatrix} q^{\frac{1}{2}}a & q^{-\frac{1}{2}}b \\ q^{\frac{1}{2}}c & q^{-\frac{1}{2}}d \end{pmatrix}X,$$

$$Y\begin{pmatrix} a & b \\ c & d \end{pmatrix} = \begin{pmatrix} q^{-\frac{1}{2}}aY + (q^{\frac{1}{2}} - q^{-\frac{3}{2}})bX & q^{\frac{1}{2}}bY \\ q^{-\frac{1}{2}}cY + (q^{\frac{1}{2}} - q^{-\frac{3}{2}})dX & q^{\frac{1}{2}}dY \end{pmatrix}$$

$$\Delta X = X \otimes ag + Y \otimes cg + 1 \otimes X, \quad \Delta Y = X \otimes bg + Y \otimes dg + 1 \otimes Y$$

as well as $\Delta g = g \otimes g$ and the matrix comultiplication for the matrix of generators.

Proof The coproduct is the right-handed cross product by the coaction of $\widetilde{SL_q(2)}$ on $\mathbb{A}_q^2$. We then compute the induced action via the above functor and make the corresponding right-handed cross product. □

Moreover, any braided group B coacts on itself by its coproduct Δ. Hence by the (right comodule version of the) bosonisation theorem, there is a canonical right coaction of $H {\ltimes} B$ on B itself. In the case of the quantum plane, we obtain a coaction of $\widetilde{SL_q(2)} {\ltimes} \mathbb{A}_q^2$ on $\mathbb{A}_q^2$ given by

$$\beta(x, y) = (x, y) \otimes 1 + 1 \otimes (X, Y) + (x, y) \otimes \begin{pmatrix} a & b \\ c & d \end{pmatrix} g.$$

Thus we identify this quantum group as the inhomogeneous quantum group of special linear transformations, scale transformations and translations of the quantum plane, in other words, the quantum group of affine transformations of the quantum plane. In higher dimensions, this can be used for example for the construction of natural q-Poincaré (+scale) quantum groups. Note that when $q \neq 1$, we cannot avoid the extension by $k_{\lambda^{-1}}\mathbb{Z}$.

17

Double bosonisation. Diagrammatic construction of $u_q(sl_2)$

Whereas the last lecture was about 'bosonisation' and led to the $u_q(b_+)$ quantum groups, this time we will do 'double bosonisation' and the construction of $u_q(sl_2)$ by these braided group methods. Note that this approach is a little idiosyncratic but it provides, in my opinion, the deepest insight into the construction of quantum groups of this type. This is also the last lecture with 'braid diagrams' – we finish up this middle section of the course. We will, however, link this approach with Lusztig's in the lecture after this one.

First of all, a few final remarks about bosonisation. Although we gave the proof quite explicitly, its real content is no more than the (quite elementary) construction of a monoidal category from B – the category $_B\mathcal{C}$ of braided modules in $\mathcal{C} = {}_H\mathcal{M}$. After this, it is more or less clear that there is some ordinary Hopf algebra (in our case $B{>\!\!\!\triangleleft}H$) whose modules are this category. The abstract formulation of this last step is called 'Tannaka–Krein reconstruction' – given a monoidal category $\mathcal{C}$ and a monoidal functor $\mathcal{C} \to \mathrm{Vec}$ obeying some technical conditions, there is an ordinary Hopf algebra in Vec such that the functor factors through its modules. In nice cases this identifies $\mathcal{C}$ with this category of modules. There is also a technically superior comodule version.

Moreover, bosonisation itself provides a kind of 'functor' from braided groups of a certain kind (in module or comodule categories) to ordinary Hopf algebras. There is also a partially defined 'functor' in the other direction.

Proposition 17.1 *Every quasitriangular Hopf algebra $H, \mathcal{R}$ has a braided group version B (called its* transmutation*) consisting of*

    1. $B = H$ as an algebra.

    2. B an H-module under Ad.

3. *The coproduct*

$$\Delta_B b = \sum b_{(1)} S\mathcal{R}^{(2)} \otimes \mathrm{Ad}_{\mathcal{R}^{(1)}}(b_{(2)}), \quad \forall b \in B$$

and the counit of H. *The right hand side here is in terms of the Hopf algebra structure of* H.

Proof The conceptual reason is as follows. The Tannaka–Krein reconstruction mentioned above can be generalised to a monoidal functor $\mathcal{C} \to \mathcal{V}$ between monoidal categories, in which case it of course reconstructs a braided group in $\mathcal{V}$ rather than in Vec. Then, one can essentially take the identity functor ${}_H\mathcal{M} \to {}_H\mathcal{M}$ and reconstruct this time a braided group in ${}_H\mathcal{M}$ (rather than recovering H). On the other hand, one may verify the statement directly by elementary (but quite long) computations. For example, we already know that H lives in ${}_H\mathcal{M}$ as an algebra under Ad. □

There is also a dual version, of course. If H is a dual quasitriangular Hopf algebra then it has a braided version B given by
1. $B = H$ as a coalgebra.
2. B an object in $\mathcal{M}^H$ under the adjoint coaction Ad_R.
3. The modified product

$$h \cdot g = \sum h_{(2)} g_{(2)} \mathcal{R}((Sh_{(1)})h_{(3)} \otimes Sg_{(1)}), \quad \forall h, g \in H$$

in terms of the product of H. We already seen that H as a coalgebra lives in $\mathcal{M}^H$ via Ad_R. As an example, the transmutation of $SL_q(2)$ is $BSL_q(2)$ as described in Example 15.6. Its product is obtained by this construction and it is for this reason that it is covariant under Ad_R and a braided group.

A couple of remarks: if H is a finite-dimensional quasitriangular Hopf algebra, let B^* be the braided group associated to H^*. Its bosonisation turns out to be

$$B^* {\rtimes} H \cong D(H),$$

the quantum double. In other words, the complicated quantum double construction is in this case actually isomorphic to a cross product as an algebra and as a coalgebra!

Also when H is quasitriangular, if B is the braided group associated to H and B^* is associated to H^*, the map $Q : H^* \to H$ defined by $\mathcal{R}_{21}\mathcal{R}$ is a homomorphism

$$B^* \to B$$

of braided groups. Since B^* is also the dual of B, it means that when H is factorisable, the braided versions of H and H^* are actually isomorphic as braided groups and self-dual. So they are more like $\mathbb{R}^n$ than nonAbelian groups! For example, $BSL_q(2)$ and the braided version $BU_q(sl_2)$ are essentially the same (infinite-dimensional) braided group, even though one looks very much like a 'matrix coordinate ring' and the other very much like an enveloping algebra. This self-duality is only possible when $q \neq 1$ and accounts for some of the most remarkable features of this class of quantum groups.

We turn now to this lecture's main topic. We start with a diagrammatic construction.

Proposition 17.2 *Let B' be dually paired with braided group B. There is a monoidal category $_{B'}C_B$ of crossed bimodules consisting of objects $(V, \triangleright, \triangleleft)$ where*

1. *V is a braided right B-module under $\triangleleft$.*
2. *V is a braided left B'-module under $\triangleright$.*
3. *The compatibility condition in Figure 17.1(a) holds.*

Morphisms are morphisms in C commuting with the actions of B, B'. The tensor product is the braided tensor product of these modules. Moreover, when B is rigid then $_{B^*}C_B$ is a braided monoidal category with Ψ as given in Figure 17.1(b).

Proof The proof that the tensor product actions of B and B' on $V \otimes W$ again obey the compatibility condition 3 is shown in Figure 17.1(c). That the braiding defined in the case were there is also a coevaluation $\text{coev} = \frown$ indeed obeys the required hexagon conditions is shown in part (d). $\qquad\square$

This is a braided version of the braided category of crossed modules $\mathcal{M}_H^H$ except that we rework an H-comodule as an H'-module where H' is dually paired to H (and then pass to the braided version). So all the ideas here are familiar by now, but we do it diagrammatically in some braided category. Let us also note that this category also comes with two monoidal functors,

$$_{B'}C \longleftarrow _{B'}C_B \longrightarrow C_B$$

which forget respectively the B action and the B' action. Here C_B is the monoidal category of right B-modules in C, etc.

In order to apply this construction concretely we need to 'break' the

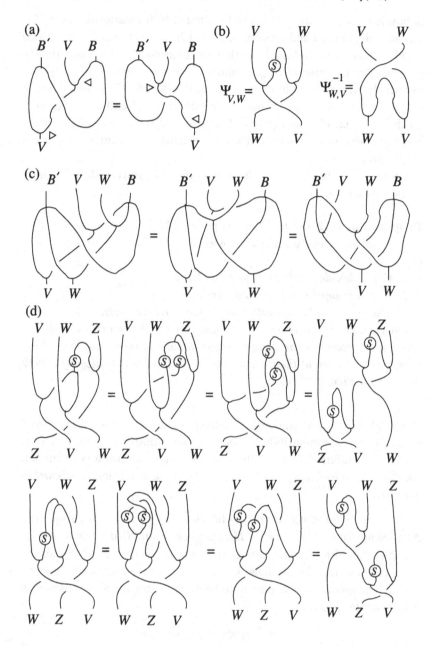

Fig. 17.1. Braided crossed bimodules $_{B'}\mathcal{C}_B$.

left–right symmetry by choosing to represent the crossed bimodule category as either left or right modules of some ordinary Hopf algebra. We chose left modules. Then in our concrete setting where $\mathcal{C} = {}_H\mathcal{M}$ we know from Theorem 16.4 that $_{B'}\mathcal{C}$ is equivalent to modules of the bosonisation $B' {>\!\!\!\triangleleft} H$.

Lemma 17.3 *Let* $(H, \mathcal{R})$ *be a quasitriangular Hopf algebra,* $\mathcal{C} = {}_H\mathcal{M}$ *and* $B \in \mathcal{C}$ *a braided group. Then there is an ordinary Hopf algebra* $H {\triangleright\!\!\!<} B^{\mathrm{op}}$ *such that its left module category can be identified with the braided right modules* $\mathcal{C}_B$*. It contains* B^{op} *as a subalgebra,* H *as a sub-Hopf algebra with*

$$hb = \sum (h_{(2)} {\triangleright} b) h_{(1)}, \quad \Delta b = \sum \mathcal{R}^{(2)} {\triangleright} b_{(1)} \otimes b_{(2)} \mathcal{R}^{(1)}, \quad \forall b \in B, \quad h \in H.$$

Proof This is a version of Theorem 16.4, and proven similarly. We have as an algebra $B^{\mathrm{op}} \in \mathcal{M}_H$, where the right action of H is $b {\triangleleft} h = (Sh) {\triangleright} b$, so we make a right cross product $H {\triangleright\!\!\!<} B^{\mathrm{op}}$ by this. An object $V \in \mathcal{C}_B$ means a right action of B (which we identify with a left action of B^{op}) and a left action of H, and the action of B being a morphism corresponds to the cross relations stated. The tensor product of braided right B-modules (reflect Figure 16.2 and restore the braid crossings) is

$$(v \otimes w) {\triangleleft} b = v {\triangleleft} \Psi(w \otimes b_{(1)}) {\triangleleft} b_{(2)} = v {\triangleleft} (\mathcal{R}^{(2)} {\triangleright} b_{(1)}) \otimes (\mathcal{R}^{(1)} {\triangleright} w) {\triangleleft} b_{(2)}.$$

This coinciding with the usual tensor product of $H {\triangleright\!\!\!<} B$-modules fixes the coproduct of the bosonisation. □

Note that this is not exactly a right-handed version of Theorem 16.4 but a variant of that. We can now 'glue' these left and right bosonisations together.

Theorem 17.4 *Let* H *be quasitriangular and* $B \in \mathcal{M}_H$ *a braided group with dually paired braided group* B' *(with invertible antipode). There is an ordinary Hopf algebra* $B' {>\!\!\!\triangleleft} H {\triangleright\!\!\!<} B^{\mathrm{op}}$ *(the* double bosonisation*) given by*

1. $B' \otimes H \otimes B$ *as a vector space.*
2. *Containing* $H {\triangleright\!\!\!<} B^{\mathrm{op}}$ *as a sub-Hopf algebra.*
3. *Containing* $B' {>\!\!\!\triangleleft} H$ *as a sub-Hopf algebra*
4. *The cross relations*

$$\sum c_{(1)} \mathcal{R}^{(2)} b_{(1)} \mathrm{ev}(\mathcal{R}^{(1)} {\triangleright} c_{(2)} \otimes b_{(2)}) = \sum \mathrm{ev}(c_{(1)} \otimes \mathcal{R}^{(2)} {\triangleright} b_{(1)}) b_{(2)} \mathcal{R}^{(1)} c_{(2)}$$

for all $b \in B^{\mathrm{op}}$ *and* $c \in B'$.

Moreover, the right modules are in 1-1 correspondence with crossed bimodules in $_{B'}\mathcal{C}_B$ (an identification of monoidal categories).

Proof We know that we have ordinary Hopf algebras $H{\bowtie}B^{\mathrm{op}}$ by the above and $B'{\rtimes}H$ by the last lecture. We embed these Hopf algebras as $1 \otimes H{\bowtie}B^{\mathrm{op}}$ and $B'{\rtimes}H \otimes 1$ and then specify the relations between $c = c \otimes 1 \otimes 1$ and $b = 1 \otimes 1 \otimes b$ to define the full algebra structure. One may then verify that the coproducts extend to a coproduct on the whole algebra by a direct computation if one wants (it is worth noting that the expressions in the cross relations 4 are just the bosonised coproducts as already given).

The correspondence with braided crossed bimodules is

$$(c \otimes h \otimes b){\triangleright}v = c{\triangleright}(h{\triangleright}(v{\triangleleft}b)), \quad \forall c \in B, \ h \in H, \ b \in B,$$

where the actions on the right hand side are the left action of B' and the right action of B, and the left action of H on V as an object in $_H\mathcal{M}$. We check that the B', B^{op} cross relations correspond correctly to the crossed bimodule compatibility condition. Thus, reading the left hand side of Figure 17.1(a), the specified map is

$$c \otimes v \otimes b \mapsto c_{(1)} \otimes c_{(2)} \otimes v{\triangleleft}b_{(1)} \otimes b_{(2)}$$
$$\mapsto c_{(1)} \otimes \mathcal{R}^{(2)}{\triangleright}(v{\triangleleft}b_{(1)}) \otimes \mathcal{R}^{(1)}{\triangleright}c_{(2)} \otimes b_{(2)}$$
$$\mapsto c_{(1)}{\triangleright}(\mathcal{R}^{(2)}{\triangleright}(v{\triangleleft}b_{(1)}))\,\mathrm{ev}(\mathcal{R}^{(1)}{\triangleright}c_{(2)} \otimes b_{(2)})$$

which corresponds to the left hand side of the cross relation 4. Similarly for the other side of the cross relations. With more care one can obtain the explicit product on the tensor product of the three vector spaces. $\square$

The inclusion of the two sub-Hopf algebras in the double bosonisation corresponds to the two partially forgetful functors above, which become pull-backs under the stated identification. Moreover,

Corollary 17.5 *When B is rigid and $B' = B^*$, then the double bosonisation is quasitriangular, with quasitriangular structure*

$$\mathcal{R}.\exp^{-1}, \quad \exp^{-1} = \sum_a Se_a \otimes f^a \in B \otimes B^*$$

where $\{e_a\}$ is a basis of B and $\{f^a\}$ a dual basis. The element $\exp^{-1}$ and $\mathcal{R} \in H \otimes H$ are viewed in $(B^*{\rtimes}H{\bowtie}B^{\mathrm{op}})^{\otimes 2}$.

Proof On general grounds we know that a braiding in the category of modules corresponds precisely in the present context to a quasitriangular structure. Comparing its induced braiding with the one from Figure 17.1(a) gives the formula shown. The latter sends

$$v \otimes w \mapsto \sum_a v \triangleleft Se_a \otimes f^a \triangleright w \mapsto \sum_a \mathcal{R}^{(2)} \triangleright (f^a \triangleright w) \otimes \mathcal{R}^{(1)} \triangleright (v \triangleleft Se_a)$$

which we identify as the braiding corresponding to $\mathcal{R} \exp^{-1}$ viewed as an element of $(B' {\rtimes} H {\ltimes} B^{\mathrm{op}})^{\otimes 2}$. Here coev $= \frown = \sum_a e_a \otimes f^a$ and is denoted exp in the context of (braided) Fourier theory. $\square$

Note that when $H = k$, a braided group in $\mathcal{M}_H$ means an ordinary Hopf algebra and the above theorem reduces to the quantum double $D(H)$ which we have given previously as corresponding to crossed modules. However, we can now construct $u_q(sl_2)$ as the simplest possible braided example.

Example 17.6 *Let q be a primitive odd n'th root of 1. View $B = k[E]/\langle E^n \rangle$ as a braided group in the category of left $k_{q^2}\mathbb{Z}/_n$-modules by $g \triangleright E = q^2 E$. Its double bosonisation is $u_q(sl_2)$.*

Proof By similar computations to those in the last lecture, we find that $H {\ltimes} B^{\mathrm{op}}$ is precisely $u_q(b_+) \subset u_q(sl_2)$. We identify $B^* = k[F]/\langle F^n \rangle$, where

$$\mathrm{ev}(F \otimes E) = \frac{1}{q^{-1} - q}$$

(this is an arbitrary normalisation which we chose) and $g \triangleright F = q^{-2} F$ as the dual braided group (another copy of the reduced braided line). We already know from the last lecture that its bosonisation is $u_q(b_-)$. Here

$$\mathcal{R}^{(2)} \otimes \mathcal{R}^{(1)} \triangleright F = g^{-1} \otimes F, \quad \mathcal{R}^{(2)} \triangleright E \otimes \mathcal{R}^{(1)} = E \otimes g$$

from the form of $\mathcal{R}$ for $k_{q^2}\mathbb{Z}/_n$ in Example 6.4. The new part is the cross relations

$$FE + g^{-1} \mathrm{ev}(F \otimes E) = \mathrm{ev}(F \otimes E)g + EF$$

from Theorem 17.4 (given the additive coproducts of the braided groups) to give the correct cross relations of $u_q(sl_2)$ as in Lecture 6. Finally, for the quasitriangular structure we have basis $e_a = E^a$ and hence

$$f^a = \frac{E^a}{[a]_{q^2}!}(q^{-1} - q)^a, \quad \exp^{-1} = e_{q^{-2}}^{E \otimes F(q - q^{-1})}$$

by absorbing the factor $q^{a(a-1)}$ from $SE^a = (-E)^a q^{a(a-1)}$ into the change of $[a]_{q^2}!$ to $[a]_{q^{-2}}!$. So we obtain the full structure of $u_q(sl_2)$ as built by double bosonisation of the reduced braided line. $\square$

Let us remark also that the above double bosonisation theorem and its corollary allow for induction. Given a quasitriangular Hopf algebra we can classify all possible braided groups in its category of modules. For each finite-dimensional one we obtain a new quasitriangular Hopf algebra. Thus,

$$k_{q^2}\mathbb{Z}_{/n} \xrightarrow{\mathrm{A}_q^1} u_q(sl_2) \xrightarrow{\mathrm{A}_q^2} u_q(sl_3) \xrightarrow{\mathrm{A}_q^3} u_q(sl_4)$$

etc. where we use as above reduced finite-dimensional versions of A_q^m at q a root of unity. Of course, we are not really limited to roots of unity; there are similar formulations of the above over formal powerseries and in other contexts as well. Moreover, from the node $U_q(sl_2)$ (say) there are in fact several interesting braided groups in its category of modules. As well as A_q^2, we have a 3-dimensional quantum plane corresponding to the identification with $U_q(so_3)$ (in the formal powerseries setting), which leads to $U_q(so_5)$. In this way, we can construct the q-deformations of all complex semisimple Lie algebras inductively, as the quantum group version of the idea of adding a node to a Dynkin diagram. Moreover, there are many more braided groups than the q-deformations of the standard geometrical planes, leading to a larger 'quantum Dynkin classification' for all the possible branches at each node of the induction, largely unexplored.

18

The braided group $U_q(n_+)$. Construction of $U_q(\mathfrak{g})$

In this lecture we are finally able to construct $U_q(\mathfrak{g})$ for general complex semisimple Lie algebras $\mathfrak{g}$, as an example of the general double-bosonisation construction. We have already indicated how they may be built up by repeatedly adjoining quantum-braided planes via the double bosonisation – in effect we accumulate all these steps into a single braided group B and define $U_q(\mathfrak{g}) = B' {\rtimes} H {\ltimes} B^{\mathrm{op}}$. This is much more useful than simply giving a large number of generators and relations to check: for example it implies by construction a 'triangular decomposition' of $U_q(\mathfrak{g})$.

Before doing this, we will later need one more general result about double bosonisation, which there was not space to include in the last lecture.

Proposition 18.1 *Let B be a braided group in a braided category $\mathcal{C}$ and dually paired with B'. Assume that B, B' have invertible antipode. Then the category ${}_{B'}\mathcal{C}_B$ has a canonical object $(V, \triangleright, \triangleleft)$ given by*

1. $V = B$ as an object of $\mathcal{C}$.

2. $\triangleleft$ the right adjoint action of B on B as shown in the upper box in Figure 18.1.

3. $\triangleright$ the right coregular action of B' on B made into a left action, as shown in the lower box in Figure 18.1.

The product of B becomes a morphism in ${}_{B'}\mathcal{C}_B$.

Proof This is proven by diagrams as shown in Figure 18.1. We show that $\triangleright, \triangleleft$ obey the crossed bimodule compatibility property of ${}_{B'}\mathcal{C}_B$, using the elementary properties of braided groups from Lectures 14 and 15. Each of these actions makes B a braided module algebra – by similar

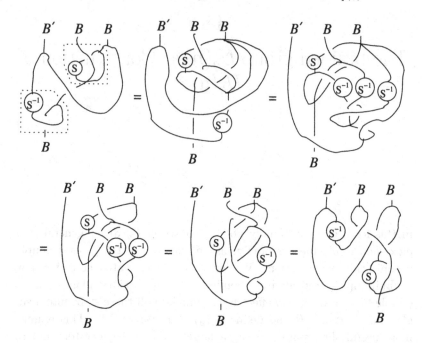

Fig. 18.1. Canonical object of ${}_{B'}C_B$ by adjoint and coregular actions.

diagrammatic computations as in Lectures 14 and 15, i.e. the product of B is a morphism in the crossed bimodule category.   □

Corollary 18.2 *Let $B, B' \in {}_H\mathcal{M}$ in the setting of the double boson-isation theorem. Then B is a $B' {>\!\!\!\triangleleft} H {\bowtie} B^{\mathrm{op}}$-module algebra under the canonical action*

$$b {\triangleright} v = \sum S(\mathcal{R}^{(2)} {\triangleright} b_{(1)}) \, (\mathcal{R}^{(1)} {\triangleright} v) \, b_{(2)},$$

$$c {\triangleright} v = \sum \mathrm{ev}(S^{-1}(\mathcal{R}^{(1)} {\triangleright} c) \otimes v_{(2)}) \, \mathcal{R}^{(2)} {\triangleright} v_{(1)}, \quad \forall v \in B, \; b \in B^{\mathrm{op}}, \; c \in B'$$

where the braided group structures on the right are those of B. H itself acts on B as an object in ${}_H\mathcal{M}$.

Proof Here $V = B$ as above. The corollary follows by the correspondence between the objects of ${}_{B'}C_B$ and left modules of $B' {>\!\!\!\triangleleft} H {\bowtie} B^{\mathrm{op}}$. We read the explicit formulae off the diagrammatic definitions in Figure 18.1.   □

The action of B' here is basically braided differentiation as we have seen in Lecture 15, while the action of Ad is typically related to conformal transformations.

Example 18.3 *Let q be a primitive n'th root of unity. $k[x]/\langle x^n \rangle$ is a $u_q(sl_2)$-module algebra. The action on any $f(x)$ is*

$$E \triangleright f = qx^2 \partial_{q^2} f, \quad F \triangleright f = -\partial_{q^{-2}} f, \quad g \triangleright f = f(q^2 x).$$

Proof We let $x = (q^{-1} - q)E$ so that $\mathrm{ev}(F \otimes x) = 1$ (recall the normalisation of the pairing in the construction of $u_q(sl_2)$ in the last lecture). By similar computations as there, we easily obtain $E \triangleright f = -E(g \triangleright f) + fE = (f(x)x - xf(q^2 x))/(q^{-1} - q)$ which can be written as stated. Also $F \triangleright f = -\mathrm{ev}(F \otimes f_{(2)})g^{-1} \triangleright f_{(1)}$ which is again a q-derivative. The same result clearly holds for generic q. $\qquad\square$

We now begin the construction of the general $u_q(\mathfrak{g})$, or more precisely $U_q(\mathfrak{g})$ (not requiring q a root of unity). This is more or less Lusztig's construction albeit in a modern braided category interpretation. Thus, following Lusztig:

Definition 18.4 *A Cartan datum is a set $I = \{i\}$ and a symmetric bilinear form $\cdot$ on $\mathbb{Z}[I]$ such that*

$$i \cdot i \in \{2, 4, 6, \dots\}, \quad a_{ij} \equiv \frac{2i \cdot j}{i \cdot i} \in \{0, -1, -2, \dots\}, \quad \forall i \neq j.$$

For those who know about semisimple Lie algebras, a_{ij} is a symmetrisable Cartan matrix. From this we build a dual quasitriangular Hopf algebra

$$H' = k\mathbb{Z}[I] = k[g_i, g_i^{-1}]_{i \in I}, \quad \mathcal{R}(g_i \otimes g_j) = q^{i \cdot j}.$$

Here $\mathbb{Z}[I]$ is the free Abelian group on I and we take its group algebra. We also need the following additional structure:

Definition 18.5 *A root datum is the additional structure (on a Cartan datum) consisting of*

1. Two finitely generated free Abelian groups Y, X with a nondegenerate pairing $\langle \, , \, \rangle : Y \times X \to \mathbb{Z}$.

2. Inclusions (the second denoted $i \mapsto i'$)

$$Y \supseteq I \subseteq X, \quad \text{such that} \quad \langle i, j' \rangle = a_{ij}.$$

In this case we define a Hopf algebra

$$H = kY = \{K_\mu \mid \mu \in Y\}, \quad \Delta K_\mu = K_\mu \otimes K_\mu, \quad K_\mu K_\nu = K_{\mu+\nu}.$$

Here H is the group algebra of the group Y but we have written a basis of elements explicitly, labelled by $\mu \in Y$. We also have a duality pairing

$$\langle g_i, K_\mu \rangle = q^{\langle \mu, i' \rangle}$$

with H', where $q \in k^*$ is fixed. In Lusztig's setting $k = \mathbb{Q}(q)$ but we can work more generally. We extend the inclusion $I \subseteq X$ to a group homomorphism $\mathbb{Z}[I] \to X$ and pull back the pairing to $\mathbb{Z}[I] \times X \to \mathbb{Z}$. This defines the pairing between the corresponding group Hopf algebras with the help of q.

Since H' is dual quasitriangular, we would like H to be quasitriangular. This is not literally true, but there is a weaker structure. Recall from Problem I.14 that if H is a finite-dimensional Hopf algebra then $\mathcal{R}$ can be viewed as an algebra and anticoalgebra map $H^* \to H$. We also have the conjugate quasitriangular structure $\bar{\mathcal{R}} = \mathcal{R}_{21}^{-1}$ which can also be viewed as a map $\bar{\mathcal{R}} : H^* \to H$. This provides the following third formulation of Drinfeld's axioms, this time for a dual pair of Hopf algebras.

Definition 18.6 A weak quasitriangular structure *on dually paired Hopf algebras H, H' is*

1. Convolution-invertible algebra and anticoalgebra maps $\mathcal{R}, \bar{\mathcal{R}} : H' \to H$.

2. $\langle \psi, \bar{\mathcal{R}}(\phi) \rangle = \langle \phi, \mathcal{R}^{-1}(\psi) \rangle$ for all $\phi, \psi \in H'$, where $\mathcal{R}^{-1}$ denotes the convolution inverse.

3.

$$R^*_{\cdot}(h) = \mathcal{R} * L^*_{\cdot}(h) * \mathcal{R}^{-1} = \bar{\mathcal{R}} * L^*_{\cdot}(h) * \bar{\mathcal{R}}^{-1}, \quad \forall h \in H,$$

where $*$ is the convolution product in $\mathrm{Hom}(H', H)$ and the coregular actions R^*, L^* are viewed as maps $H' \to H$ for each $h \in H$.

Recall that the convolution product is defined for maps from any coalgebra to any algebra (see Lemma 8.3).

Proposition 18.7 *A choice of Cartan and root data precisely provide a weakly quasitriangular dual pair $H' = k\mathbb{Z}[I]$ and $H = kY$, with*

$$\mathcal{R}(g_i) = K_i^{\frac{i \cdot i}{2}}, \quad \bar{\mathcal{R}}(g_i) = K_i^{-\frac{i \cdot i}{2}}.$$

Proof We clearly have well-defined algebra homomorphisms $\mathcal{R}, \bar{\mathcal{R}}$ as stated. Moreover, $\mathcal{R}^{-1} = \bar{\mathcal{R}}$ as a consequence of the symmetry of $\cdot$. We check that $\langle g_i, \mathcal{R}^{-1}(g_j) \rangle = q^{-\frac{i \cdot i}{2} \langle j, i' \rangle} = q^{-j \cdot i} = q^{-i \cdot j} = q^{-\frac{i \cdot i}{2} \langle i, j' \rangle} = \langle g_j, K_i^{-\frac{i \cdot i}{2}} \rangle = \langle g_j, \bar{\mathcal{R}}(g_i) \rangle$ again by the symmetry of $\cdot$. $\square$

Clearly, there is a straightforward generalisation of the double bosonisation theorem to this weakly quasitriangular setting. Where we had a left action of H we replace it by evaluation against a right coaction of H'. Where we had $\mathcal{R}^{(1)} \triangleright v \otimes \mathcal{R}^{(2)}$ we replace it by $v^{(\bar{1})} \otimes \mathcal{R}(v^{(\bar{2})})$. Where we had $\mathcal{R}^{-(1)} \otimes \mathcal{R}^{-(2)} \triangleright v$ we replace it by $\bar{\mathcal{R}}(v^{(\bar{2})}) \otimes v^{(\bar{1})}$. With this generalisation, we can double-bosonise any braided group in $\mathcal{M}^{H'}$ to obtain a Hopf algebra $B \bowtie H \ltimes B^{\mathrm{op}}$.

In our case, we now let $\tilde{B} = k\langle e^i \rangle_{i \in I}$ be the free noncommutative algebra on I. This lives in the braided category of right H'-comodules by $e^i \mapsto e^i \otimes g_i$. Hence it also lives in the category of left H-modules by

$$K_\mu \triangleright e^i = \langle g_i, K_\mu \rangle e^i = q^{\langle \mu, i' \rangle} e^i.$$

In the category of right H'-comodules, we extend

$$\Delta e^i = e^i \otimes 1 + 1 \otimes e^i, \quad \epsilon e^i = 0, \quad S e^i = -e^i$$

as a braided group, in just the same way as we have seen for the braided line $\mathbb{A}_q^1$. We have a $\mathbb{Z}[I]$-grading corresponding to the H'-coaction in place of the $\mathbb{Z}$-grading as for the braided line, but otherwise the construction is analogous. The braiding from the dual quasitriangular structure and the coaction is

$$\Psi(e^i \otimes e^j) = e_j \otimes e_i \mathcal{R}(g_i \otimes g_j) = q^{i \cdot j} e_j \otimes e_i.$$

Similarly, we take $\tilde{B}' = k\langle f_i \rangle_{i \in I}$ the free algebra again, with right coaction $f_i \mapsto f_i \otimes g_i^{-1}$. It has the additive braided group structure as above, and the braiding $\Psi(f_i \otimes f_j) = f_j \otimes f_i q^{i \cdot j}$ in a similar way. We take it as dually paired by

$$\mathrm{ev}(f_i \otimes e^j) = \delta_i^j (q_i^{-1} - q_i)^{-1}, \quad q_i = q^{\frac{i \cdot i}{2}}$$

and we assume that $q_i^2 \neq 1$ for all $i \in I$.

Finally, the duality paring between $\tilde{B}$ and $\tilde{B}'$ is actually very degenerate. We make it nondegenerate by dividing through by the kernels (the coradicals) of the pairing,

$$\ker_2 \mathrm{ev} = \{ b \in \tilde{B} \mid \mathrm{ev}(c \otimes b) = 0, \ \forall c \in \tilde{B}' \}$$

and similarly for $\ker_1$ ev. It is obvious that

$$B' = \tilde{B}'/\ker_1 \text{ev}, \quad B = \tilde{B}/\ker_2 \text{ev}$$

remain braided groups with the inherited coactions, braidings etc., now nondegenerately paired. (This is also a useful idea for ordinary Hopf algebra pairings, by the way.) One can check that the kernels are ideals from the axioms of a duality pairing, and by the same axioms that the coproducts descend. Since ev is a morphism in the category of comodules, the quotients remain in this category as well.

Example 18.8 *Given a Cartan and root datum and $H, B, \tilde{B}$ as above, and $q \in k^*$ such that $q_i^2 \neq 1$ for all i, we have an ordinary Hopf algebra $U_q = B' {>\!\!\!\triangleleft} H {\triangleright\!\!\!<} B^{\mathrm{op}}$. It consists of generators e_i, f_i, K_μ, the relations*

$$K_\mu K_\nu = K_{\mu+\nu}, \quad K_\mu e^i = q^{\langle \mu, i' \rangle} e^i K_\mu, \quad f_i K_\mu = q^{\langle \mu, i' \rangle} K_\mu f_i,$$

$$[e^i, f_j] = \frac{K_i^{\frac{i \cdot i}{2}} - K_i^{-\frac{i \cdot i}{2}}}{q_i - q_i^{-1}} \delta_j^i$$

and the relations coming from the kernels of ev. *The coalgebra is*

$$\Delta e^i = e^i \otimes K_i^{\frac{i \cdot i}{2}} + 1 \otimes e^i, \quad \Delta f_i = f_i \otimes 1 + K_i^{-\frac{i \cdot i}{2}} \otimes f_i,$$

$$\epsilon(K_\mu) = 1, \quad \epsilon(e^i) = \epsilon(f_i) = 0.$$

The antipode has the usual form uniquely determined by the above.

Proof The relations $hb = (h_{(2)} {\triangleright} b) h_{(1)}$ and $ch = h_{(1)} (c {\triangleleft} h_{(2)})$ in $H {\triangleright\!\!\!<} B^{\mathrm{op}}$ and $B' {>\!\!\!\triangleleft} H$ give the relations with K_μ as stated. By the slight generalisation to the weak case (as explained), the coproducts of the bosonisations likewise come out as stated. Finally, from the weak version

$$\sum c_{(1)} \mathcal{R}(c_{(2)}{}^{(\bar{2})}) b_{(1)} \text{ev}(c_{(2)}{}^{(\bar{1})} \otimes b_{(2)}) = \sum \text{ev}(c_{(1)}{}^{(\bar{2})} \otimes b_{(1)}) b_{(2)} \bar{\mathcal{R}}(c_{(1)}{}^{(\bar{1})}) c_{(2)}$$

for all $b \in B^{\mathrm{op}}$ and $c \in B'$, of the cross relations in the double bosonisation theorem, we have $f_j e^i + \mathcal{R}(g_i^{-1}) \text{ev}(f_j \otimes e^i) = \bar{\mathcal{R}}(g_j^{-1}) \text{ev}(f_j \otimes e^i) + e^i f_j$, which comes out as stated. The computations are basically the same as for $u_q(sl_2)$, proceeding directly from the additive form of the coproducts of $B, \tilde{B}$ on the generators. □

In the case where the Cartan matrix corresponds to a complex semisimple Lie algebra $\mathfrak{g}$ (or even a Kac–Moody one), we write $U_q = U_q(\mathfrak{g})$ with subalgebras $B' = U_q(n_-)$ and $B^{\mathrm{op}} = U_q(n_+)$. Unlike Lie theory, however, we also use the root datum, which has the result that we distinguish for example between $U_q(su_2)$ and $U_q(so_3)$ – one is the double cover of the other. Here U_q has for its Cartan part the group torus rather than the enveloping algebra of the Cartan Lie subalgebra. This is the version of $U_q(\mathfrak{g})$ due to Lusztig, albeit in a modern setting of braided groups and in conventions with the opposite coproduct. We also have a quantum group

$$\tilde{U}_q = \tilde{B}' {>\!\!\!\triangleleft} H {\triangleright\!\!\!<} \tilde{B}^{\mathrm{op}}$$

in the same way, without making the quotients needed to make the pairing nondegenerate. In the next lecture we will see how to compute these quotients more explicitly.

19

q-Serre relations

In the last lecture we constructed the quantum group U_q as the double bosonisation of the braided group $B = U_q^+$ (say), defined as the quotient of $\tilde{B}$ by the kernel of the pairing ev with $\tilde{B}'$. Let us similarly denote by U_q^- the braided group B' obtained from the quotient of $\tilde{B}'$ by the kernel of the other input of the pairing. When the Cartan and root data correspond to a Lie algebra $\mathfrak{g}$ we have

$$U_q^+ = U_q(n_+)^{\mathrm{op}}, \quad U_q^- = U_q(n_-), \quad U_q = U_q(n_-) {\rtimes} H {\ltimes} U_q(n_+) = U_q(\mathfrak{g})$$

where H is a 'maximal torus' group Hopf algebra. The construction is, however, more general.

We now study these braided groups $U_q^\pm$ in much more detail, in particular describing the kernel of ev which defines them. By the theory in previous lectures we know that the modules of U_q are the crossed U_q^-–U_q^+-braided bimodules. In the same way, most of the properties of U_q are reduced to properties of these braided groups.

We recall that

$$\tilde{B} = k\langle e^i \rangle_{i \in I}, \quad \Delta e^i = e^i \otimes 1 + 1 \otimes e^i, \quad \epsilon(e^i) = 0, \quad S e^i = -e^i$$

is the free algebra and forms a braided group with 'coaddition' as shown. Similarly, $\tilde{B}' = k\langle f_i \rangle_{i \in I}$. For the present purposes we do not need the strange normalisation that we used before, so we will work with

$$y_i = f_i(q_i^{-1} - q_i), \quad \mathrm{ev}(y_i \otimes e^j) = \delta_i^j$$

where the pairing is more conventional. These are braided groups in the category of $\mathbb{Z}[I]$-graded spaces, or $k\mathbb{Z}[I]$-comodules, where

$$e^i \mapsto e^i \otimes g_i, \quad y_i \mapsto y_i \otimes g_i^{-1}, \quad \mathcal{R}(g_i \otimes g_j) = q^{i \cdot j}$$

are the coactions and dual quasitriangular structure of $k\mathbb{Z}[I]$. It implies the braidings

$$\Psi(e^i \otimes e^j) = q^{i \cdot j} e^j \otimes e^i, \quad \Psi(e^i \otimes y_j) = q^{-i \cdot j} y_j \otimes e^i,$$

etc. This is the only part that we will need now. On the other hand, we have already studied many similar braided groups, such as the braided line. Similarly, we define $\partial_i : \tilde{B} \to \tilde{B}$ by

$$\partial_i f = \text{coeff}_{e^i \otimes} \Delta f, \quad \forall f \in \tilde{B}.$$

This is the evaluation of y_i with the first factor of Δf.

Lemma 19.1 ∂_i *obeys*

$$\partial_i e^j = \delta_i^j, \quad \partial_i(fg) = (\partial_i f)g + q^{i \cdot |f|} f \partial_i g, \quad \forall f, g \in \tilde{B},$$

where f is of homogeneous $\mathbb{Z}[I]$-degree $|f|$. Here

$$|(e^{i_1})^{n_1} \dots (e^{i_m})^{n_m}| = n_1 i_1 + \dots + n_m i_m \in \mathbb{Z}[I].$$

Proof The braided-Leibniz rule is obtained from the braided module algebra property of the coregular action in the category with opposite braiding, as in Lecture 15. In our case $\Psi^{-1}(y_i \otimes e^j) = q^{i \cdot j} e^j \otimes y_i$. Note also that $q^{i \cdot i} = q_i^2$, so that

$$\partial_i(e^i)^n = [n]_{q_i^2}(e^i)^{n-1},$$

which is just as for the braided line.   □

Moreover, we can express the evaluation ev in terms of these differentials.

Lemma 19.2 *For all $i_1, \dots, i_n$ and all n,*

$$\text{ev}(y_{i_1} y_{i_2} \dots y_{i_n} \otimes f) = \quad \text{} \quad = \epsilon \circ \partial_{i_1} \partial_{i_2} \dots \partial_{i_n} f, \quad \forall f \in \tilde{B}.$$

Proof This follows from the property of a braided pairing – and the diagrammatic definition of ∂_i.   □

This is the braided-geometrical meaning of the quotient from $\tilde{B}$ to B; we set to zero the functions that have zero differentials. Note that for usual polynomial algebras, or algebras like them, there are no such functions other than the constant ones; this is not so when the algebras are free.

For technical reasons we will actually prefer to work with $\bar{\partial}_i : \tilde{B} \to \tilde{B}$ obeying

$$\bar{\partial}_i e^j = \delta_i^j, \quad \bar{\partial}_i(fg) = (\bar{\partial}_i f)g + q^{-i \cdot |f|} f \bar{\partial}_i g, \quad \forall f, g \in \tilde{B}$$

with f of homogeneous degree. This has just the same form as ∂_i but with q replaced by q^{-1}. As we saw in Example 18.3, $-\bar{\partial}_i$ is the coregular action defined by evaluation against the right output of the coproduct (the lower box in Figure 18.1) and has the conceptual advantage that it defines a braided module algebra in our original category (not the opposite one). It is immediate from their diagrammatic definitions that $S \circ \partial_i = -\bar{\partial}_i \circ S$ where S is the braided group antipode. Hence we can equally well write

$$\ker_2 \text{ev} = \{ f \in \tilde{B} \mid \epsilon \circ \bar{\partial}_{i_1} \ldots \bar{\partial}_{i_n} f = 0, \quad \forall i_1, \ldots, i_n, \ \forall n \}$$

(because the kernel is clearly S-invariant).

We are now ready to describe the kernel of ev defining U_q^+. In the literature, this is usually given as generated by

$$(*) \qquad \sum_{r=0}^{1-a_{ij}} (-1)^r \begin{bmatrix} 1 - a_{ij} \\ r \end{bmatrix}_{q_i} (e^i)^r e^j (e^i)^{1-a_{ij}-r}, \quad \forall i \neq j,$$

the setting of which to zero are the 'q-Serre relations'. The q-binomials here use the symmetric q-integer conventions (see the end of Lecture 7). Similarly for U_q^- on the other side of the pairing.

Proposition 19.3 *The expression* $(*)$ *is* $\text{Ad}_{(e^i)^{1-a_{ij}}}(e^j)$, *where* Ad *is the right braided adjoint action of* $\tilde{B}$ *on itself (the upper box in Figure 18.1).*

Proof Reading from the diagrammatic definition of Ad in the upper box in Figure 18.1, we have

$$\text{Ad}_{(e^i)^n}(e^j) = \sum_{r=0}^{n} (-1)^r q_i^{r(r-1)} (e_i)^r q_i^{r a_{ij}} \begin{bmatrix} n \\ r \end{bmatrix}_{q_i^2} (e^j)(e^i)^{n-r}.$$

We use the coproduct $\Delta(e_i)^n$ as for the braided line (since $\Psi(e^i \otimes e^i) = q_i^2 e^i \otimes e^i$), and likewise the antipode is similar to that of the braided line. We also use $\Psi(e^j \otimes (e^i)^r) = q^{r \cdot i \cdot j}(e^i)^r \otimes e^j$ and $q^{i \cdot j} = q_i^{a_{ij}}$ from the definition of a_{ij}. To compare the result with $(*)$, we note that

$$\begin{bmatrix} n \\ r \end{bmatrix}_q = q^{-nr}q^{r^2} \begin{bmatrix} n \\ r \end{bmatrix}_{q^2}$$

relates the two conventions that we use for q-integers and associated factorials etc. We also need $n = 1 - a_{ij}$. □

On the other hand, from the last lecture we know that $-\bar{\partial}_i$ and Ad fit together to make $\tilde{B}$ a braided crossed $\tilde{B}'$–$\tilde{B}$ bimodule. Hence, essentially without computation, we have:

Corollary 19.4

$$\bar{\partial}_j \mathrm{Ad}_{e^i}(f) = \mathrm{Ad}_{e^i}\bar{\partial}_j(f) + \delta_j^i(q^{i \cdot |f|} - q^{-i \cdot |f|})f$$

for all $f \in \tilde{B}$ of homogeneous degree.

Proof We have basically seen this in the last lecture. The coregular action in the lower of the two boxes in Figure 18.1 is $y_i \triangleright f = -\bar{\partial}_i f$ as mentioned above (or one can just compute it by reading off the diagram). Meanwhile, the $\mathrm{Ad}_{e^i}, -\bar{\partial}_i$ relations can be read off from the crossed bimodule condition Figure 17.1(a) (again using the linear form of the coproducts). Equivalently, they represent the cross relations of $\tilde{B}' \rtimes H \ltimes \tilde{B}^{\mathrm{op}}$. □

From this, we see easily that

$$\bar{\partial}_k \mathrm{Ad}_{(e^i)^{1-a_{ij}}}(e^j) = 0, \quad \forall k, \ \forall i \neq j.$$

The case $k \neq i$ is particularly easy as then $\bar{\partial}_k$ and Ad_{e^i} commute. The case $k = i$ requires a little more care because we accumulate terms proportional to

$$q_i[-a_{ij}] + q_i[-a_{ij} - 1] + \cdots + q_i[a_{ij}],$$

which vanishes by symmetry. Here

$$|\mathrm{Ad}_{(e^i)^r}(e_j)| = ri + j, \quad q^{i \cdot (ri+j)} = q_i^{2r+a_{ij}}.$$

Hence,

$$\mathrm{Ad}_{(e_i)^{1-a_{ij}}}(e^j) \in \ker_2 \mathrm{ev}, \quad \forall i \neq j$$

as claimed. This is the braided-geometric meaning of the q-Serre relations of U_q^+.

Finally, it can be shown that for Cartan data corresponding to complex semisimple or Kac–Moody Lie algebras, and regular choices of root data, these elements actually generate the whole of $\ker_2$ ev. It is beyond our scope to prove this (Lusztig quotes results from the representation theory of Kac–Moody Lie algebras to prove it), but assuming it, we have an explicit description of $U_q(n_+)$ as $k\langle e^i \rangle$ modulo the relations

$$\mathrm{Ad}_{(e_i)^{1-a_{ij}}}(e^j) = 0, \quad \forall i \neq j.$$

Similarly for $U_q(n_-)$. This gives in turn the fully explicit description of $U_q(\mathfrak{g})$ by generators and relations. It is also possible (Drinfeld) to simply define $U_q(\mathfrak{g})$ with such generators and relations and verify that they form a Hopf algebra by hand.

In passing, we can give the explicit form of the action of U_q corresponding to the above braided crossed bimodule.

Corollary 19.5 U_q^+ *is a U_q-module algebra by*

$$K_\mu \triangleright f = q^{\langle \mu, |f|' \rangle} f, \quad e^i \triangleright f = \frac{f x^i - q^{i \cdot |f|} x^i f}{q_i^{-1} - q_i}, \quad f_i \triangleright f = -\bar{\partial}_i f, \quad \forall f \in U_q^+.$$

Proof This generalises Example 18.3, and the computation is basically the same. We let $x_i = (q_i^{-1} - q_i)e_i$ so that $\mathrm{ev}(f_i \otimes x^j) = 1$ and use these for U_q^+. We also have the action of $H = kY$ corresponding to U_q^+ as an object in the category of $k\mathbb{Z}[I]$-comodules. $\quad\square$

This is a kind of 'Verma module' action of $U_q(\mathfrak{g})$ on $U_q(n_+)^{\mathrm{op}}$ but in our case as a module algebra. It also works at the level of $\tilde{U}_q$, where it is another way to think about the q-Serre relations as invariant elements. Also, $\mathrm{Ad}_{(e^i)^r}(e^j)$ for $r < 1 - a_{ij}$ generate other elements of $U_q(n_+)$ and from this and the triangular decomposition, one can obtain a 'Poincaré–Birkhoff–Witt' (PBW)-like basis along similar lines to the PBW basis of $U(\mathfrak{g})$. We also know from the general diagrammatic constructions that $U_q(\mathfrak{g})$ has some kind of quasitriangular structure, e.g. in the finite-dimensional versions at roots of unity, or by working with suitable topological duals. Clearly, one can go on and develop many more of the features of $U_q(\mathfrak{g})$. For example, there is a Lusztig–Kashiwara 'canonical basis' of $U_q^\pm$ with many remarkable properties and characterised up to sign by equations involving the $\bar{\partial}_i$.

Finally, to justify our notations, we say a few words about the usual enveloping algebra $U(\mathfrak{g})$ and its deformation in the powerseries setting. Thus, let $\mathfrak{g}$ be a complex semisimple Lie algebra (say) and $\mathfrak{t}$ a Cartan

Lie subalgebra. Then $\mathfrak{t}^*$ has on it a symmetric bilinear form $(\ ,\ )$ defined by the inverse of the Killing form. Finally, fix a system of positive roots and let $\alpha_i \in \mathfrak{t}^*$ be the simple roots. We have the integers

$$d_i = (\alpha_i, \alpha_i)/2, \quad a_{ij} = \frac{2(\alpha_i, \alpha_j)}{(\alpha_i, \alpha_i)}.$$

On the other hand, the Lie algebra $\mathfrak{g}$ is spanned by $H_i \in \mathfrak{t}$ and root vectors $X_{\pm\alpha}$, which in turn are generated by repeated applications of the Lie algebra adjoint action $\mathrm{ad}_{X_{\pm i}} = [X_{\pm i}, \ ]$ from simple root vectors $X_{\pm j}$. Hence we can consider $H_i, X_{\pm i}$ alone as the generators of $U(\mathfrak{g})$, in which case this is the free algebra generated by these modulo the relations

$$[H_i, H_j] = 0, \quad [H_i, X_{\pm j}] = \pm a_{ij} X_{\pm j},$$

$$[X_{+i}, X_{-j}] = \delta_{ij} H_i, \quad \mathrm{ad}_{X_{\pm i}}^{1-a_{ij}}(X_{\pm j}) = 0$$

(the last for $i \neq j$; the Serre relations).

We see that the structure of $U_q(\mathfrak{g})$ is a deformation of $U(\mathfrak{g})$ when we write $q = e^{\frac{t}{2}}$, $K_i = q^{H_i}$ and define E_i, F_i from $X_{\pm i}$ (and $q^{\pm\frac{H_i}{2}}$), cf. the $U_q(sl_2)$ case. In this setting there is a quasitriangular structure

$$\mathcal{R} = q^{\sum_{i,j} d_i (a^{-1})_{ij} H_i \otimes H_j} \left(\prod_{\alpha > 0} e_{q_\alpha^{-2}}^{(q_\alpha - q_\alpha^{-1}) E_\alpha \otimes F_\alpha} \right),$$

where $q_\alpha = q^{d_\alpha}$ and $d_\alpha = (\alpha, \alpha)/2$. The general E_α, F_α are defined by repeated applications of the braided adjoint action Ad as above. The exact specification of this, as well as the specification of ordering on the set of positive roots in $\prod_{\alpha>0}$, are both fixed by a choice of maximal element of the Weyl group of the root system.

20

R-matrix methods

This lecture finishes up our study of quasitriangular Hopf algebras and related topics with some general R-matrix methods, as promised earlier. We will sketch, in particular, the construction of matrix quantum groups G_q dual to the quantum groups $U_q(\mathfrak{g})$.

Recall that if $t^i{}_j$ are a matrix of generators of a dual quasitriangular Hopf algebra $(H, \mathcal{R})$ (with a matrix form of coalgebra structure, as for $SL_q(2)$) then $R^i{}_j{}^k{}_l = \mathcal{R}(t^i{}_j \otimes t^k{}_l)$ is a matrix $R \in M_n \otimes M_n$ obeying the Yang–Baxter equations. We also have

$$R^{-1}{}^i{}_j{}^k{}_l = \mathcal{R}(St^i{}_j \otimes t^k{}_l), \quad \tilde{R}^i{}_j{}^k{}_l = \mathcal{R}(t^i{}_j \otimes St^k{}_l),$$

where $\tilde{R} = ((R^{t_2})^{-1})^{t_2}$ (t_2 transposition in the second factor of M_n) is called the *second inverse* of R.

Conversely, let $R \in M_n \otimes M_n$ obey the Yang–Baxter relations and be biinvertible in the sense that $R^{-1}, \tilde{R}$ exist. We have already claimed that one can define a dual quasitriangular bialgebra $A(R)$ as $k\langle t^i{}_j \rangle$ modulo the relations $Rt_1 t_2 = t_2 t_1 R$, and will indicate a proof of the dual quasi-triangularity now. We use here a compact notation where $t \in M_n(A(R))$ is a matrix of generators and $t_1 = t \otimes \mathrm{id} \in (M_n \otimes M_n)(A(R))$. Similarly, $t_2 = \mathrm{id} \otimes t$; the numerical suffixes denote the copy of M_n and matrix multiplication is understood. The relations of $A(R)$ were written more explicitly in Lecture 8.

Lemma 20.1 *The bialgebra $A(R)$ has two canonical representations $\rho^\pm$ in M_n, defined by $\rho_2^+(t_1) = R$ and $\rho_2^-(t_1) = R_{21}^{-1}$, or more explicitly*

$$\rho^+(t^i{}_j)^k{}_l = R^i{}_j{}^k{}_l, \quad \rho^-(t^i{}_j)^k{}_l = R^{-1}{}^k{}_l{}^i{}_j.$$

Proof It is easy to see in the compact notation that the relations are indeed represented. Thus,

$$\rho_3^+(R_{12}\mathbf{t}_1\mathbf{t}_2) = R_{12}\rho_3^+(\mathbf{t}_1)\rho_3^+(\mathbf{t}_2) = R_{12}R_{13}R_{23} = R_{23}R_{13}R_{12}$$
$$= \rho_3^+(\mathbf{t}_2)\rho_3^+(\mathbf{t}_1)R_{12} = \rho_3^+(\mathbf{t}_2\mathbf{t}_1R_{12}),$$

using the Yang–Baxter equations. Similarly,

$$\rho_3^-(R_{12}\mathbf{t}_1\mathbf{t}_2) = R_{12}\rho_3^-(\mathbf{t}_1)\rho_3^-(\mathbf{t}_2) = R_{12}R_{31}^{-1}R_{32}^{-1} = R_{32}^{-1}R_{31}^{-1}R_{12}$$
$$= \rho_3^-(\mathbf{t}_2)\rho_3^-(\mathbf{t}_1)R_{12} = \rho_3^-(\mathbf{t}_2\mathbf{t}_1R_{12}),$$

using $R_{12}R_{31}^{-1}R_{32}^{-1} = R_{32}^{-1}R_{31}^{-1}R_{12}$, i.e. $R_{31}R_{32}R_{12} = R_{12}R_{32}R_{31}$, which is again the Yang–Baxter equations after a relabelling of the positions in $M_n^{\otimes 3}$. $\qquad\square$

We have given this first proof in full detail as a demonstration of the methods. The basic idea is to use the compact matrix notation and the Yang–Baxter equations over and over again. Using such R-matrix methods,

Theorem 20.2 *The bialgebra $A(R)$ is dual quasitriangular as defined by $\mathcal{R}(\mathbf{t}_1 \otimes \mathbf{t}_2) = R$.*

Proof We sketch this only (it is long but fairly straightforward). Again, suffixes refer to positions in a tensor product of matrices M_n. We have to show that $\mathcal{R}$ extends to products of the generators in a manner consistent with the relations of $A(R)$. First of all, $\mathcal{R}((\ ) \otimes \mathbf{t}) = \rho^+$ so this extension is the lemma above. Since $A(R)$ is a bialgebra with $\Delta\mathbf{t} = \mathbf{t} \otimes \mathbf{t}$, we have tensor product representations $\rho^{+ \otimes m}$ which provide the values of $\mathcal{R}((\ ) \otimes \mathbf{t}_1\mathbf{t}_2\ldots\mathbf{t}_m)$. Arguing similarly for the antirepresentation $\mathcal{R}(\mathbf{t} \otimes (\ ))$ gives $\mathcal{R}(\mathbf{t}_1\ldots\mathbf{t}_{m'} \otimes(\ ))$. In this way one obtains the same $\mathcal{R}$ (given on general products in both inputs by an array of R-matrices) as a well-defined map $A(R) \otimes A(R) \to k$. Using ρ^- we likewise construct $\mathcal{R}^{-1}$. The dual quasitriangularity axioms then hold by construction and by the relations of $A(R)$. $\qquad\square$

For example, when $R = R_{sl_2}$ (as obtained from the 2-dimensional representation of $U_q(sl_2)$ in Lecture 11), we have $A(R) = M_q(2)$. This is therefore the natural way to encode the six q-commutativity relations of that bialgebra. We also know that we can quotient by $\det_q(\mathbf{t}) = 1$ to obtain a dual quasitriangular Hopf algebra $SL_q(2)$, or we can invert $\det_q(\mathbf{t})$ and obtain $GL_q(2)$ as another dual quasitriangular Hopf algebra. The biinvertibility of R is a necessary condition for either of these constructions. (It is also sufficient for the existence of a certain maximal Hopf algebra built from $A(R)$.)

More generally all the linear algebraic groups $k[G]$ of the type associated to complex semisimple $\mathfrak{g}$ have standard q-deformations G_q defined as quotients of $A(R)$ for suitable R, which are known for all $q \in k^*$ possessing certain roots. This constructs them as dual quasitriangular Hopf algebras. For example,

$$R_{sl_n} = q^{-\frac{1}{n}} \left(q \sum_i E_{ii} \otimes E_{ii} + \sum_{i \neq j} E_{ii} \otimes E_{jj} + (q - q^{-1}) \sum_{j > i} E_{ij} \otimes E_{ji} \right)$$

defines $SL_q(n)$, where E_{ij} is the matrix which is 1 at row i and column j and zero elsewhere. It is the bialgebra $A(R)$ modulo certain q-determinant relations. For the B, C, D series one has similar R-matrices and additional relations q-deforming the classical characterisation of the groups as subsets of M_n. For the exceptional groups one has candidates for suitable R-matrices, but the explicit form of the exceptional Hopf algebras G_q is not known at the moment.

Similarly, there is an R-matrix formulation for quantum or 'braided' planes. The data here are $R, R' \in M_n \otimes M_n$ where

1. R is a biinvertible solution of the Yang–Baxter equations.
2. R' obeys $(\Psi + 1)(\Psi' - 1) = 0$ and $\Psi'\Psi = \Psi\Psi'$, where $\Psi = \tau \circ R$ and $\Psi' = \tau \circ R'$.
3. $R'_{12}R_{13}R_{23} = R_{23}R_{13}R'_{12}$, $R_{12}R_{13}R'_{23} = R'_{23}R_{13}R_{12}$.

In fact, the simplest way to satisfy 3 is to assume that $\Psi' = f(\Psi)$ for some polynomial f. Then 3 becomes empty. For simplicity, we specialise to this case now. It is also easy to solve 2. Namely, any R has some minimal polynomial for the corresponding Ψ. Writing $\{\lambda_i\}$ for the (possibly repeated) roots of this, we can suppose w.l.o.g. that one of them is $\lambda_i = -1$ (by choice of normalisation of R). Then

$$\Psi' = \prod_{j \neq i} (\Psi - \lambda_j \mathrm{id}) + \mathrm{id}.$$

Proposition 20.3 For R, R' as above, there is a braided group $V(R', R)$ in the category $\mathcal{M}^{A(R)}$, defined as $k\langle x_i \rangle$ modulo the relations

$$x_i x_j = \sum_{a,b} x_b x_a R'^a{}_i{}^b{}_j,$$

with the 'additive' coalgebra, coaction and induced braiding

$$\Delta x_i = x_i \otimes 1 + 1 \otimes x_i, \quad \epsilon(x_i) = 0, \quad S x_i = -x_i,$$

$$x_i \mapsto \sum_a x_a \otimes t^a{}_i, \quad \Psi(x_i \otimes x_j) = \sum_{a,b} x_b \otimes x_a R^a{}_i{}^b{}_j.$$

Proof This most easily proven by R-matrix methods in the compact notation $\mathbf{x}_1\mathbf{x}_2 = \mathbf{x}_2\mathbf{x}_1 R'$ (where the suffixes denote the placement of the covector $\mathbf{x} = (x_i)$ against the indicated copy of M_n). It is easy to check that $V(R', R)$ is indeed a right $A(R)$-comodule algebra, which requires condition 3. It implies the braiding $\Psi(\mathbf{x}_1 \otimes \mathbf{x}_2) = \mathbf{x}_2 \otimes \mathbf{x}_1 R$. We then check that Δ extends as a braided group coproduct with this braiding. Thus

$$\begin{aligned}
\Delta(\mathbf{x}_1\mathbf{x}_2) &= (\mathbf{x}_1 \otimes 1 + 1 \otimes \mathbf{x}_1)(\mathbf{x}_2 \otimes 1 + 1 \otimes \mathbf{x}_2) \\
&= \mathbf{x}_1\mathbf{x}_2 \otimes 1 + 1 \otimes \mathbf{x}_1\mathbf{x}_2 + \mathbf{x}_1 \otimes \mathbf{x}_2 + \Psi(\mathbf{x}_1 \otimes \mathbf{x}_2)
\end{aligned}$$

where the product is in $V(R', R) \underline{\otimes} V(R', R)$. Similarly for $\Delta \mathbf{x}_2\mathbf{x}_1 R'$. Equality of these requires condition 2. $\square$

By the above, we have at least as many possible such quantum planes for a given R as the eigenvalues of Ψ. Solutions with just two distinct eigenvalues $q, -q^{-1}$ are called *q-Hecke*. The entire R_{sl_n} family are of this type when normalised suitably; they obey $(\Psi - q)(\Psi + q^{-1}) = 0$ as we have seen for R_{sl_2} in Lecture 11.

Example 20.4 *The choice*

$$R = q^{\frac{3}{2}} R_{sl_2}, \quad R' = q^{-\frac{1}{2}} R_{sl_2}$$

recovers the quantum plane $\mathbb{A}_q^2 \equiv \mathbb{A}_q^{2|0}$ as a braided group. On the other hand, the choice

$$R = -q^{-\frac{1}{2}} R_{sl_2}, \quad R' = -q^{\frac{3}{2}} R_{sl_2}$$

defines a finite-dimensional 'fermionic' quantum plane $\mathbb{A}_q^{0|2}$. When $q^2 \neq 1$, this is $k\langle x, y \rangle$ modulo the relations

$$x^2 = 0, \quad y^2 = 0, \quad yx = -q^{-1}xy,$$

the additive braided coalgebra structure on the generators and the braiding

$$\Psi(x \otimes x) = -x \otimes x, \quad \Psi(y \otimes y) = -y \otimes y, \quad \Psi(x \otimes y) = -q^{-1}y \otimes x$$

$$\Psi(y \otimes x) = -q^{-1}x \otimes y + (q^{-2} - 1)y \otimes x.$$

Proof We use R_{sl_2} from the R_{sl_n} family above (or as given in Lecture 11). The combinations R, R' we use do not actually need q to have a square root, in fact the resulting matrices involve only powers of q^2. [Also, the term 'fermionic' comes from physics – when $q = 1$, we obtain a 2-dimensional affine superplane.] $\qquad\square$

Each of these braided groups $V(R', R)$ lives in a braided category and we can proceed as in previous lectures to define braided partial derivatives $\partial_i : V(R', R) \to V(R', R)$ as the coefficient of $x_i \otimes$ id in Δf. [Explicitly,

$$\partial^i x_{i_1} \dots x_{i_m} = \delta^i{}_{j_1} x_{j_2} \dots x_{j_m} \, [m; R]^{j_1 \dots j_m}_{i_1 \dots i_m}$$

where

$$[m; R] = 1 + \Psi_{12} + \Psi_{12}\Psi_{23} + \dots + \Psi_{12} \dots \Psi_{m-1,m} \in (M_n)^{\otimes m}$$

is a *braided integer* matrix. It generalises $1 + q + \dots + q^{m-1} = [m]_q$ to the general case. We have braided factorial and binomial matrices, etc. as well.] We also have vectorial quantum-braided planes $V^*(R', R)$, and R-matrix formulae for the matrix braided groups $B(R)$ related to $A(R)$ by transmutation.

Finally, we consider briefly the problem of classifying the different possible solutions R of the Yang–Baxter equations. This has been done by computer in lower dimensions, over $\mathbb{C}$ and up to some natural symmetries:

(i) For $n = 3$ (9×9 matrices) all upper triangular invertible solutions have been classified. This includes the standard R_{sl_3} and R_{so_3} families.

(ii) For $n = 2$ (4×4 matrices) all invertible solutions have been classified. There are eight parameterised families, of which one is the standard R_{sl_2} family. For example, a totally different but still q-Hecke 'eight-vertex' family is

$$\frac{1}{2}\begin{pmatrix} q - q^{-1} + 2 & 0 & 0 & q - q^{-1} \\ 0 & q + q^{-1} & q - q^{-1} & 0 \\ 0 & q - q^{-1} & q + q^{-1} & 0 \\ q - q^{-1} & 0 & 0 & q - q^{-1} - 2 \end{pmatrix}.$$

Of the eight families, one has three eigenvalues, one has four and the rest have two eigenvalues. There are also three isolated solutions.

The 'group theory' associated to the nonstandard solutions R is little explored. Moreover, in general, very little is known about this 'Yang–Baxter variety' (where we can consider $R^i{}_j{}^k{}_l$ as indeterminates in $k^{n^2 \times n^2}$, subject to the cubic Yang–Baxter relation). Its detailed structure remains little explored.

On the other hand, one point is always the identity matrix $1 \in M_n \otimes M_n$. Passing through this point are the standard deformations such as R_{sl_n} or R_{so_n} etc. – this tangent space at 1 is more or less fully understood and corresponds essentially to deformations of Lie algebras and solutions closely related to them.

More generally, we have a 'bundle' of quantum groups, braided groups and in fact of entire braided geometries over each point R in the Yang–Baxter variety; classical Lie theory appears only as the tangent space at 1. (We will say more about this in Lecture 22). Similarly, the theory of super-Lie algebras appears as the tangent space over different trivial solutions (which are like 1 but with some minus signs). Moreover, there are other solutions R not connected to either of these points. For example, the twist map τ is always an isolated solution (it is not biinvertible, however).

Also, we note that based on what happens at 1, one might try to understand the tangent to the Yang–Baxter variety at a generic point R_0 in terms of braided-Lie algebras and braided geometry in the category of $A(R_0)$-comodules that lives over the point R_0. If this could be done uniformly, one should then be able to integrate to obtain flows of solutions. Again, these ideas remain little explored at the moment.

21

Group, algebra, Hopf algebra factorisations. Bicrossproducts

This lecture is devoted to a completely different class of quantum groups associated to group factorisations and generalisations of them. I do not want to leave the impression that $U_q(\mathfrak{g})$ are the only noncommutative and noncocommutative Hopf algebras of wide interest.

Definition 21.1 *A* group factorisation *is a group X and subgroups $G, M \subseteq X$ such that the product map provides a set bijection $G \times M \cong X$.*

The more familiar case is where G, say, is normal. Then X is a group semidirect product. A general group factorisation does not require this, however. Instead:

Proposition 21.2 *If X is a group factorisation into G, M then M acts from the left on the set of G and G acts from the right on the set of M, by actions $\triangleright, \triangleleft$ obeying*

$$e \triangleleft u = e, \quad (st) \triangleleft u = (s \triangleleft (t \triangleright u))\,(t \triangleleft u), \quad s \triangleright e = e, \quad s \triangleright (uv) = (s \triangleright u)\,((s \triangleleft u) \triangleright v)$$

for all $u, v \in G$ and $s, t \in M$. (A pair of groups acting on each other in this way is called a matched pair*.) Moreover, $X \cong G \bowtie M$, where the latter is*

1. *The set $G \times M$.*

2. *The product and inverse $(u, s)(v, t) = (u(s \triangleright v), (s \triangleleft v)t)$, $(u, s)^{-1} = (s^{-1} \triangleright u^{-1}, s^{-1} \triangleleft u^{-1})$, and unit (e, e).*

Conversely, given a matched pair of groups, we have a group factorisation $X = G \bowtie M$, called the double cross product.

Proof Consider $su \in X$ where $s \in M, u \in G$. By the unique factorisation assumption, this element is equal to $(s \triangleright u)(s \triangleleft u)$ for some elements $s \triangleright u \in$

G and $s \triangleleft u \in M$. This defines the maps $\triangleright : M \times G \to G$ and $\triangleleft : M \times G \to G$. One checks from associativity of X that $s \triangleleft e = s$, $(s \triangleleft u) \triangleleft v = s \triangleleft (uv)$ and $e \triangleright u = u$, $s \triangleright (t \triangleright u) = (st) \triangleright u$ (i.e. that these are group actions) and then that the matched pair conditions hold. In the converse direction it is a nice exercise to verify associativity and the inverse for $G \bowtie M$ as defined. $\qquad\square$

The idea in this lecture is that whereas conventional mathematics considers direct and semidirect products, the much more natural concept is a more symmetrical one where each acts on the other. [The motivation from physics is the general principle that every action has a 'backreaction'.] In fact, group factorisations abound in group theory although not usually thought of in terms of mutual actions like this. The construction is, however, classical in origin. More recently, (unital) algebras have been considered in the same way.

Definition 21.3 *An algebra factorisation is an algebra X and two subalgebras $A, B \subseteq X$ such that the product map defines a linear isomorphism $A \otimes B \cong X$.*

Again, algebra factorisations abound. For example, one has $M_n(\mathbb{C}) = (\mathbb{C}\mathbb{Z}_{/n})(\mathbb{C}\mathbb{Z}_{/n})$ as a factorisation of $n \times n$ matrices into group algebras. A general class of nontrivial examples which we have already seen in Lecture 14 is the braided tensor product $A \underline{\otimes} B$ of H-module algebras (where H is quasitriangular).

Proposition 21.4 *If X is an algebra factorisation into A, B then there is a linear map $\Psi : B \otimes A \to A \otimes B$ (the 'generalised braiding') obeying*

$$\Psi \circ (m \otimes \mathrm{id}) = (\mathrm{id} \otimes m) \circ \Psi_{12} \circ \Psi_{23}, \quad \Psi(1 \otimes a) = a \otimes 1, \quad \forall a \in A$$

$$\Psi \circ (\mathrm{id} \otimes m) = (m \otimes \mathrm{id}) \circ \Psi_{23} \circ \Psi_{12}, \quad \Psi(b \otimes 1) = 1 \otimes b, \quad \forall b \in B,$$

where the suffixes refer to the tensor factor on which Ψ acts and m denotes the relevant product map. Moreover, $X \cong A \underline{\otimes} B$, where the latter is

1. *$A \otimes B$ as a vector space.*
2. *The product $(a \otimes b)(c \otimes d) = a\Psi(b \otimes c)d$ for all $a, c \in A$ and $b, d \in B$. The unit is $1 \otimes 1$.*

Conversely, given algebras A, B and map Ψ as above, we have an algebra $A \underline{\otimes} B$ defined in this way.

Proof By the factorisation, $ba \in X$ has the form $m \circ \Psi(b \otimes a)$ for some element $\Psi(b \otimes a) \in A \otimes B$ (and the product in X), for all $a \in A$, $b \in B$. We then deduce the conditions on Ψ as stated. Actually, we have seen the proofs here in the diagrammatic construction of the braided tensor product algebra $A\underline{\otimes}B$ in Lecture 14. In the proof there, we did not use entirely that Ψ is a braiding but only that it was functorial under the products maps of A, B, which are the conditions on Ψ shown. $\qquad\square$

This does not look too much like the group case, but it does when the algebras are counital (equipped with characters $\epsilon : X \to k$). This happens in the bialgebra or Hopf algebra case, for example. Thus,

Definition 21.5 *A bialgebra factorisation is a bialgebra X and subbialgebras $A, H \subseteq X$ such that the product map defines a linear isomorphism $A \otimes H \cong X$.*

We do the theory for bialgebras; the same applies for Hopf algebras without additional conditions.

Proposition 21.6 *If X is a bialgebra factorisation into bialgebras A, H then H is a right A-module coalgebra and A is a left H-module coalgebra such that*

$$(hg){\triangleleft}a = \sum(h{\triangleleft}(g_{(1)}{\triangleright}a_{(1)}))(g_{(2)}{\triangleleft}a_{(2)}), \quad 1{\triangleleft}a = \epsilon(a),$$

$$h{\triangleright}(ab) = \sum(h_{(1)}{\triangleright}a_{(1)})((h_{(2)}{\triangleleft}a_{(2)}){\triangleright}b), \quad h{\triangleright}1 = \epsilon(h),$$

$$\sum h_{(1)}{\triangleleft}a_{(1)} \otimes h_{(2)}{\triangleright}a_{(2)} = \sum h_{(2)}{\triangleleft}a_{(2)} \otimes h_{(1)}{\triangleright}a_{(1)}, \quad \forall a, b \in A, \ h, g \in H.$$

(This is called a matched pair of bialgebras.) Moreover, $X \cong A{\bowtie}H$ where the latter is

1. $A \otimes H$ as a coalgebra.

2. $(a \otimes h)(b \otimes g) = \sum a(h_{(1)}{\triangleright}b_{(1)}) \otimes (h_{(2)}{\triangleleft}b_{(2)})g$ for all $a, b \in A$ and $h, g \in H$, and unit $1 \otimes 1$.

Conversely, given a matched pair of bialgebras, $A{\bowtie}H$ is a bialgebra defined in this way (the double cross product).

Proof We have in particular an algebra factorisation, hence a map $\Psi : H \otimes A \to A \otimes H$. Using the counits, we define

$$\triangleright = (\text{id} \otimes \epsilon) \circ \Psi : H \otimes A \to A, \quad \triangleleft = (\epsilon \otimes \text{id}) \circ \Psi : H \otimes A \to H.$$

Applying $\mathrm{id} \otimes \epsilon, \epsilon \otimes \mathrm{id}$ to these tells us that $\triangleright$ is a left action and $\triangleleft$ is a right action, and

$$(hg)\triangleleft a = \cdot(h\triangleleft\Psi(g \otimes a)), \quad 1\triangleleft a = \epsilon(a),$$

$$h\triangleright(ab) = \cdot(\Psi(h \otimes a)\triangleright b), \quad h\triangleright 1 = \epsilon(h).$$

(This is the result at the level of algebras equipped with homomorphisms ϵ.) Finally, we use that the inclusions of A, B are coalgebra maps to deduce that the maps $A \otimes H \to X$ and $H \otimes A \to X$ defined by the product in X are coalgebra maps. Comparing them, we deduce that Ψ is also a coalgebra map and hence that

$$\Delta_{A \otimes H} \circ \Psi = (\Psi \otimes \Psi) \circ \Delta_{H \otimes A}, \quad (\epsilon \otimes \epsilon) \circ \Psi(h \otimes a) = \epsilon(a)\epsilon(h).$$

Now we apply $\mathrm{id} \otimes \epsilon \otimes \epsilon \otimes \mathrm{id}$ to the first of these to conclude that

$$h_{(1)}\triangleright a_{(1)} \otimes h_{(2)}\triangleleft a_{(2)} = \Psi(h \otimes a),$$

which implies the first two of the conditions for a matched pair of actions. Applying instead the map $\epsilon \otimes \mathrm{id} \otimes \mathrm{id} \otimes \epsilon$ gives

$$h_{(1)}\triangleleft a_{(1)} \otimes h_{(2)}\triangleright a_{(2)} = \tau \circ \Psi(h \otimes a),$$

where τ is the usual transposition, which proves the third condition for a matched pair. Likewise, applying instead $\epsilon \otimes \mathrm{id} \otimes \epsilon \otimes \mathrm{id}$ gives that $\triangleleft$ is a coalgebra map, while applying $\mathrm{id} \otimes \epsilon \otimes \mathrm{id} \otimes \epsilon$ gives that $\triangleright$ is a coalgebra map. In the converse direction, one verifies directly that one has a bialgebra $A\bowtie H$ from the given data. $\qquad\square$

If A, H are Hopf algebras, then so is the double cross product, with $S(ah) = (Sh)(Sa)$, and if X is a factorising Hopf algebra then its antipode is of this form.

Example 21.7 *Let H be a finite-dimensional Hopf algebra. Then the quantum double is a factorisation $D(H) = H^{*\mathrm{op}}\bowtie H$ by the mutual coadjoint actions*

$$h\triangleleft\phi = \sum h_{(2)}\langle\phi, (Sh_{(1)})h_{(3)}\rangle, \quad h\triangleright\phi = \sum \phi_{(2)}\langle h, (S\phi_{(1)})\phi_{(3)}\rangle$$

for all $h \in H$ and $\phi \in H^*$.

Proof Since $D(H)$ as given in Proposition 8.1 is clearly a Hopf algebra factorisation, we deduce this double cross product form. Computing $\triangleright, \triangleleft$ we find the mutual coadjoint actions as stated. It is a nice exercise to see directly that $H^{*\mathrm{op}}{\bowtie}H$ defined by these actions indeed recovers the structure $D(H)$ as given previously. □

Let us now turn to something a bit more conventional. Whereas a factorisation of Hopf algebras means $A \hookrightarrow X \hookleftarrow H$, we can reverse just one of these arrows and have a corresponding notion of extensions:

Definition 21.8 *An extension of bialgebras A, H is a bialgebra E and bialgebra maps*

$$0 \to A \overset{i}{\hookrightarrow} E \overset{p}{\to} H \to 0$$

such that $E{\cong}H \otimes A$ as a left H-comodule and right A-module.

An extension is strict *if the inclusion $H \hookrightarrow E$ defined by $h \mapsto 1 \otimes h$ is an algebra map, and if the surjection $E \to A$ defined by $a \otimes h \mapsto a\epsilon(h)$ is a coalgebra map.*

Here $H \otimes A$ has the obvious left coaction of H via its coproduct and the obvious right action of A by its product. Likewise, E has a coaction $(p \otimes \mathrm{id}) \circ \Delta$ and an action of A given by the pull-back of the right multiplication of E under i. The analogue of the above theorem about factorisations is then:

Proposition 21.9 *If E is a strict extension of H by A, then A is a right H-module algebra and H is a left A-comodule algebra, with some compatibility conditions between the two. Moreover, $E{\cong}H{\bowtie}A$, where the latter denotes $H{\ltimes}A$ as an algebra and $H{\rtimes}A$ as a coalgebra. (Conversely, given such a compatible action and coaction, one has a bialgebra $H{\bowtie}A$, called the* bicrossproduct.*)*

Proof This is meant to be a 'half-arrows-reversed' version of Proposition 21.6, so we omit the details. Suffice it to say that E acts on itself by the adjoint action, which we pull back to an action of H by the algebra map $H \hookrightarrow E$ in the strict case. One finds that it then restricts to an action on $A \subseteq E$. Similarly with all arrows reversed (and a left–right reversal) for the coaction of A on H. □

For Hopf algebras H, A with suitable compatible (co)actions, the resulting $H{\bowtie}A$ is automatically a Hopf algebra. Also, there is a similar theorem in the general nonstrict case, namely E is of the form $H^{\psi}{\bowtie}_{\chi}A$ where there is a cocycle $\chi : H \otimes H \to A$ and a dual cocycle $\psi : H \to A \otimes A$. (Cocycles are defined for Hopf algebras much as for

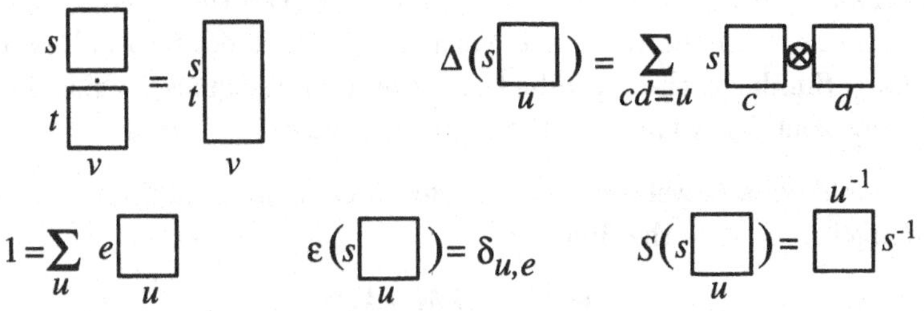

$$s\,\boxed{} \;\cdot\; \boxed{}_t \;=\; s\,\boxed{}_t \qquad\qquad \Delta\!\left(s\,\boxed{}_u\right) = \sum_{cd=u} s\,\boxed{}_c \otimes \boxed{}_d$$

$$1 = \sum_u e\,\boxed{}_u \qquad \varepsilon\!\left(s\,\boxed{}_u\right) = \delta_{u,e} \qquad S\!\left(s\,\boxed{}_u\right) = \boxed{}^{u^{-1}} s^{-1}$$

Fig. 21.1. The Hopf algebra $kM \blacktriangleright\!\!\triangleleft k(G)$.

groups.) This includes the usual theory of group extensions as a special case. Moreover, by the way we have introduced the theory above, it is clear that:

Corollary 21.10 *If A is finite dimensional, the data for a double cross product $A\bowtie H$ are in 1-1 correspondence with the data for an extension $H\blacktriangleright\!\!\triangleleft A^*$.*

Proof An action of H on A for a double cross product dualises to an action of H on A^*. An action of A on H dualises to a coaction of A^*. $\qquad\square$

Example 21.11 *If $X \cong G\bowtie M$ is a finite group factorisation, we can extend the actions linearly and hence obtain a double cross product group Hopf algebra $kG\bowtie kM$. Then by the above corollary, we have a corresponding bicrossproduct Hopf algebra $kM\blacktriangleright\!\!\triangleleft k(G)$, shown in Figure 21.1.*

Proof The left action of M on G defines a right action of kM on $k(G)$. The right action of G on M defines a left coaction of $k(G)$ on kM. The cross product algebra and coalgebra are

$$(s\otimes\delta_u)(t\otimes\delta_v) = \delta_{u,t\triangleright v}(st\otimes\delta_{uv}), \quad \Delta(s\otimes\delta_u) = \sum_{cd=u} s\otimes\delta_c \otimes s\triangleleft c\otimes\delta_d$$

for $s,t \in M$ and $u,v \in G$, where the sum is over $c,d \in G$ subject to the constraint. There is a nice diagrammatic way to describe this Hopf algebra. As basis we take labelled squares $\{s\,\boxed{}_u \mid u \in G,\ s \in M\}$. We label the top edge by $s\triangleright u$ and the right edge by $s\triangleleft u$. Then the product consists of the vertical 'glueing' of squares provided the edges match (or is otherwise zero) as shown in Figure 21.1. (We read the labelling of the glued square from the top as st, with the product nonzero only when $u = t \triangleright v$). Similarly, the coproduct is the sum of all squares which when glued horizontally would give the original square.   □

We could equally well reverse the roles of G, M and obtain a different Hopf algebra dual to the first, i.e.

$$k(M) \blacktriangleright\!\!\triangleleft kG = (kM \triangleright\!\!\blacktriangleleft k(G))^*,$$

where we take the same basis of squares and glue horizontally for the product and vertically for the coproduct. Since group factorisations abound in mathematics, so do these bicrossproduct Hopf algebras. They are noncommutative and noncocommutative, i.e. quantum groups. We give Lie group versions of them in the next lecture, but which are quite different from the $U_q(\mathfrak{g})$ examples.

22

Lie bialgebras. Lie splittings. Iwasawa decomposition

In this lecture we come back to a topic promised early on in the course. Apart from applications to knot theory, quantum groups have also had great impact on the theory of Poisson or symplectic geometry. Here we give a brief introduction to this theory, starting at the Lie algebra level. We work with general Lie algebras, but over characteristic not 2.

Definition 22.1 *A Lie bialgebra is* $(\mathfrak{g}, [\ ,\ ], \delta)$, *where*

1. $(\mathfrak{g}, [\ ,\ ])$ *is a Lie algebra.*

2. $(\mathfrak{g}, \delta)$ *is a Lie coalgebra (like a Lie algebra with arrows reversed).*

3. $\delta : \mathfrak{g} \to \mathfrak{g} \otimes \mathfrak{g}$ *is a 1-cocycle* $\delta \in Z^1_{\mathrm{ad}}(\mathfrak{g}, \mathfrak{g} \otimes \mathfrak{g})$ *in the Lie algebra cohomology. Explicitly,*

$$\delta([x, y]) = \mathrm{ad}_x(\delta(y)) - \mathrm{ad}_y(\delta(x)),$$

where $\mathrm{ad}_x(y \otimes z) = \mathrm{ad}_x(y) \otimes z + y \otimes \mathrm{ad}_x(z)$ *for all* $x, y, z \in \mathfrak{g}$.

The Lie coalgebra axiom 2 is explicitly that δ is anticocommutative and obeys the co-Jacobi identity,

$$\delta + \tau \circ \delta = 0, \quad (\delta \otimes \mathrm{id}) \circ \delta + \text{cyclic} = 0,$$

where '+cyclic' means to add the two cyclic rotations of the factors of $\mathfrak{g} \otimes \mathfrak{g} \otimes \mathfrak{g}$.

This is the infinitesimal notion of a Hopf algebra, as alluded to at the end of Lecture 7. Actually, all of Hopf algebra theory can be developed at this level, as we will now sketch.

Proposition 22.2 *If* $(\mathfrak{g}, [\ ,\ ], \delta)$ *is a finite-dimensional Lie bialgebra, then so is* $(\mathfrak{g}^*, \delta^*, [\ ,\ ]^*)$.

Proof This is a nice exercise so we omit it. The dualisation is by $\langle \phi, [x, y] \rangle = \langle \delta \phi, x \otimes y \rangle$ and $\langle [\phi, \psi], x \rangle = \langle \phi \otimes \psi, \delta x \rangle$ where $\langle \, , \, \rangle$ denotes evaluation both for $\mathfrak{g}$ and for $\mathfrak{g} \otimes \mathfrak{g}$ in the obvious way, and $x, y \in \mathfrak{g}$, $\phi, \psi \in \mathfrak{g}^*$. The only thing to check is that the axiom 3 is 'self-dual'. □

Clearly, there is a notion of dually paired infinite-dimensional Lie bialgebras as well. Similarly, notions of Lie (co)module Lie (co)algebras, etc. etc. Here we concentrate on the Lie version of Drinfeld theory.

Definition 22.3 *A quasitriangular Lie bialgebra is $(\mathfrak{g}, r)$ where*

1. $\mathfrak{g}$ is a Lie bialgebra.

2. $\delta = \partial r$ in the Lie algebra cohomology for some 0-cochain $r \in \mathfrak{g} \otimes \mathfrak{g}$. Explicitly, $\delta x = \mathrm{ad}_x(r)$.

3. $(\mathrm{id} \otimes \delta)r = [r_{13}, r_{12}]$, where the indices denote the position in $\mathfrak{g} \otimes \mathfrak{g} \otimes \mathfrak{g}$ and the Lie bracket is taken in the common (first) factor.

This is the infinitesimal notion of the axioms of a quasitriangular Hopf algebra. The motivation for this and for a Lie bialgebra come from $U_q(\mathfrak{g})$ where, in the formal powerseries setting, we have

$$\Delta - \tau \circ \Delta = t\delta + \cdots, \quad \mathcal{R} = 1 + tr + \cdots$$

to lowest order in t.

Lemma 22.4 *Let $(\mathfrak{g}, r)$ be a quasitriangular Lie bialgebra. Then*

1. $(\delta \otimes \mathrm{id})r = [r_{13}, r_{23}]$.

2. $[r_{12}, r_{13}] + [r_{12}, r_{23}] + [r_{13}, r_{23}] = 0$ (the classical Yang–Baxter equation or CYBE).

3. $\mathrm{ad}_x(r + r_{21}) = 0$ for all $x \in \mathfrak{g}$.

Proof We first check

$$(\mathrm{id} \otimes \delta)r = [r_{12}, r_{23}] + [r_{13}, r_{23}], \quad (\delta \otimes \mathrm{id})r = [r_{13}, r_{12}] + [r_{23}, r_{12}]$$

just from the definition of δ. Hence parts 1 and 2 are clear. Part 3 is just anticocommutativity of δ. □

On the other hand, to construct a quasitriangular Lie bialgebra it is enough to find $r \in \mathfrak{g} \otimes \mathfrak{g}$ which obeys the CYBE and is such that $\mathfrak{q} = r + r_{21}$ is ad-invariant. Then we can define $\delta = \partial r$ and verify from the CYBE that axiom 3 holds and that δ obeys the co-Jacobi identity (to yield a Lie coalgebra).

The element $\mathfrak{q}$ is the analogue of the element $Q = \mathcal{R}_{21}\mathcal{R}$ in the theory of quasitriangular Hopf algebras. On the other hand, being ad-invariant it is necessarily a multiple of the inverse of the Killing form when $\mathfrak{g}$ is a complex simple Lie algebra. As a result of this and other considerations, there is a complete classification (by Belavin and Drinfeld) of the possible quasitriangular Lie bialgebra structures on complex semisimple $\mathfrak{g}$. Using the notations from Lecture 19, one has in particular a standard solution

$$r = \sum_{\alpha>0} d_\alpha X_\alpha \otimes X_{-\alpha} + \frac{1}{2}\sum_{ij} d_i(a^{-1})_{ij}H_i \otimes H_j$$

called the 'Drinfeld–Sklyanin' solution of the CYBE. It is the lowest-order part of the quasitriangular structure of the standard deformations $U_q(\mathfrak{g})$, and provides a standard Lie bialgebra structure for all complex semisimple $\mathfrak{g}$.

Example 22.5 *The Lie algebra* $sl_2 = \text{span}\{H, X_\pm\}$ *with* $[H, X_\pm] = \pm 2X_\pm$ *and* $[X_+, X_-] = H$ *forms a quasitriangular Lie bialgebra by*

$$\delta(H) = 0, \quad \delta(X_\pm) = \frac{1}{2}(X_\pm \otimes H - H \otimes X_\pm), \quad r = X_+ \otimes X_- + \frac{1}{4}H \otimes H.$$

Proof It is enough to verify directly that r as stated obeys the CYBE and the ad-invariance condition, and that $\delta = \partial r$. Thus, the expression for the CYBE is

$$\frac{1}{4}[H, X_+] \otimes H \otimes X_- + \frac{1}{4}[X_+, H] \otimes X_- \otimes H$$

$$+ X_+ \otimes [X_-, X_+] \otimes X_- + \frac{1}{4}H \otimes [H, X_+] \otimes X_- + \frac{1}{4}X_+ \otimes [X_-, H] \otimes H$$

$$+ \frac{1}{4}X_+ \otimes H \otimes [X_-, H] + \frac{1}{4}H \otimes X_+ \otimes [H, X_-],$$

which vanishes by the Lie relations of sl_2. Also, we have that

$$r + r_{21} = X_+ \otimes X_- + X_- \otimes X_+ + \frac{1}{2}H \otimes H$$

is ad-invariant. ☐

This Lie bialgebra is not self-dual. Letting $\{\phi, \psi_\pm\}$ denote the dual basis to $\{H, X_\pm\}$, we find:

Example 22.6 $sl_2^* = \text{span}\{\phi, \psi_\pm\}$ is the Lie bialgebra

$$[\psi_\pm, \phi] = \tfrac{1}{2}\psi_\pm, \quad [\psi_+, \psi_-] = 0,$$

$$\delta\psi_\pm = \pm 2(\phi \otimes \psi_\pm - \psi_\pm \otimes \phi), \quad \delta\phi = \psi_+ \otimes \psi_- - \psi_- \otimes \psi_+.$$

Proof This is an exercise. We want to note mainly that whereas sl_2 is semisimple, sl_2^* is solvable. □

Finally, to complete the general theory, we have an analogue of the quantum double.

Proposition 22.7 *Let $\mathfrak{g}$ be a finite-dimensional Lie bialgebra. The classical double $D(\mathfrak{g})$ is the quasitriangular Lie bialgebra consisting of*

1. *$\mathfrak{g} \oplus \mathfrak{g}^*$ as a vector space.*
2. *$\forall x, y \in \mathfrak{g}, \ \phi, \psi \in \mathfrak{g}^*$,*

$$[\phi, \psi]_{D(\mathfrak{g})} = [\psi, \phi], \quad [x, y]_{D(\mathfrak{g})} = [x, y],$$

$$[x, \phi]_{D(\mathfrak{g})} = \sum \phi_{(1)} \langle \phi_{(2)}, x \rangle + x_{(1)} \langle \phi, x_{(2)} \rangle.$$

3. *$\delta_{D(\mathfrak{g})}\phi = \delta\phi$ and $\delta_{D(\mathfrak{g})}x = \delta x$.*
4. *$r = \sum f^a \otimes e_a$, where $\{e_a\}$ is a basis of $\mathfrak{g}$ and $\{f^a\}$ is a dual basis.*

Proof We indicate the theory behind the Lie algebra directly below. After that, it is easy to see that r obeys the CYBE and defines a Lie bialgebra. Here $\delta x = x_{(1)} \otimes x_{(2)}$ is an explicit notation for the Lie co-bracket, etc. □

To understand the Lie algebra here, we return to the factorisation ideas of the last lecture. Thus:

Definition 22.8 *A Lie algebra splitting is a Lie algebra $\mathfrak{x}$ and sub-Lie algebras $\mathfrak{g}, \mathfrak{m} \subseteq \mathfrak{x}$ such that $\mathfrak{x} \cong \mathfrak{g} \oplus \mathfrak{m}$ as a vector space by these inclusions.*

As the infinitesimal version of the result about group factorisations, we have:

Proposition 22.9 *If $\mathfrak{x}$ is a Lie algebra splitting into $\mathfrak{g}, \mathfrak{m}$ then there are Lie algebra actions*

$$\triangleleft : \mathfrak{m} \otimes \mathfrak{g} \to \mathfrak{m}, \quad \triangleright : \mathfrak{m} \otimes \mathfrak{g} \to \mathfrak{g}$$

obeying

$$\phi\triangleright[x,y] = [\phi\triangleright x, y] + [x, \phi\triangleright y] + (\phi\triangleleft x)\triangleright y - (\phi\triangleleft y)\triangleright x,$$

$$[\phi, \psi]\triangleleft x = [\phi\triangleleft x, \psi] + [\phi, \psi\triangleleft x] + \phi\triangleleft(\psi\triangleright x) - \psi\triangleleft(\phi\triangleright x)$$

for all $x, y \in \mathfrak{g}$ and $\phi, \psi \in \mathfrak{m}$ (a matched pair of Lie algebras). Moreover, $\mathfrak{x}\cong\mathfrak{g}\bowtie\mathfrak{m}$, where the latter is

1. $\mathfrak{g} \oplus \mathfrak{m}$ as a vector space.

2. $\mathfrak{m}, \mathfrak{g}$ are sub-Lie algebras and $[\phi, x] = \phi\triangleright x + \phi\triangleleft x$ for all $\phi \in \mathfrak{m}$ and $x \in \mathfrak{g}$.

Conversely, given a matched pair of Lie algebras we have a double cross sum *Lie algebra $\mathfrak{g}\bowtie\mathfrak{m}$ defined in this way.*

Proof We sketch this only. The actions $\triangleleft, \triangleright$ are obtained from $[\phi, x]$ computed in $\mathfrak{x}$ and projected respectively to $\mathfrak{g}, \mathfrak{m}$ via the splitting. Here $x \in \mathfrak{g}$ and $\phi \in \mathfrak{m}$. From the Jacobi identity one checks that these are actions

$$[\phi, \psi]\triangleright x = \phi\triangleright(\psi\triangleright x) - \psi\triangleright(\phi\triangleright x), \quad \phi\triangleleft[x, y] = (\phi\triangleleft x)\triangleleft y - (\phi\triangleleft y)\triangleleft x.$$

Further instances of the Jacobi identity provide the matched pair conditions. In the converse direction, one has to verify the Jacobi identity directly. □

One can check that $D(\mathfrak{g}) = \mathfrak{g}\bowtie\mathfrak{g}^{*\mathrm{op}}$ is a Lie algebra of this form, by the mutual coadjoint actions

$$\phi\triangleright x = \sum x_{(1)}\langle\phi, x_{(2)}\rangle, \quad \phi\triangleleft x = \sum \phi_{(1)}\langle\phi_{(2)}, x\rangle, \quad \forall x \in \mathfrak{g}, \ \phi \in \mathfrak{g}^*.$$

The splitting result above works just the same way for Lie bialgebras (with the inclusions now Lie bialgebra inclusions and $\mathfrak{g}\bowtie\mathfrak{m}$ with the direct sum Lie cobracket), in which case $D(\mathfrak{g}) = \mathfrak{g}\bowtie\mathfrak{g}^{*\mathrm{op}}$ as Lie bialgebras. We can also write it as $D(\mathfrak{g}) = \mathfrak{g}^{*\mathrm{op}}\bowtie\mathfrak{g}$ in a similar way.

As an application, recall that over $\mathbb{C}$, a real form $\mathfrak{u}$ of a Lie algebra $\mathfrak{g}$ means a basis such that the structure constants are real, i.e. a real Lie algebra whose complexification is $\mathfrak{g}$. For Lie bialgebras we define a *half-real* form as a basis where the Lie algebra structure constants are real and the Lie bialgebra structure constants are imaginary (this turns out to be more useful than the more obvious definition as both real, which would be a bialgebra over $\mathbb{R}$). In this case we define the dual half-real form $\mathfrak{u}^*$ of $\mathfrak{g}^*$ as $\imath$ times the dual basis. For example, the real form su_2 of sl_2 is a half-real form. More generally, the standard compact real form $\mathfrak{u}$ of semisimple $\mathfrak{g}$ with the Drinfeld–Sklyanin Lie bialgebra structure is a half-real form. Given a half-real form one may verify that the coadjoint actions provide mutual coadjoint actions of real Lie algebras $\mathfrak{u}, \mathfrak{u}^{\star \mathrm{op}}$ and hence a real Lie algebra $\mathfrak{u} \bowtie \mathfrak{u}^{\star \mathrm{op}}$. This is a half-real form of $D(\mathfrak{g})$. Moreover, at least for the standard Drinfeld–Sklyanin Lie bialgebra structure and standard compact real form, one finds that $\mathfrak{u} \bowtie \mathfrak{u}^{\star \mathrm{op}} \cong \mathfrak{g}$ as real Lie algebras. This is a modern proof of the Iwasawa decomposition of the complexification of $\mathfrak{u}$, into $\mathfrak{u}$ and a solvable Lie algebra which we identify as $\mathfrak{u}^{\star \mathrm{op}}$.

Finally, if G is the associated connected and simply connected Lie group to the Lie algebra $\mathfrak{g}$, and M the Lie group associated to m, then, under some technical assumptions (e.g. that G is compact) one can show that the Lie group associated to a double cross sum $\mathfrak{g} \bowtie \mathfrak{m}$ has the form $G \bowtie M$ for certain actions of G, M on each other. This is more or less obvious since the Lie groups are determined by the Lie algebras. The construction is, however, known quite explicitly, i.e. one can build the Lie group actions by exponentiating and integrating the Lie algebra actions. [It is beyond our scope to describe this here, but briefly we view the Lie actions as cocycles and exponentiate them as vector fields on the groups. From these we build a pair of zero-curvature connections on $G \times M$ viewed as a bundle over G or over M. The parallel transport under these connections provides the Lie group actions.]

Hence in particular, we obtain a modern proof of the global Iwasawa decomposition $G \cong U \bowtie U^{\star \mathrm{op}}$ where G is the complexification of the compact real form U as a real Lie group of twice the dimension. For example $SL_2(\mathbb{C}) = SU_2 \bowtie SU_2^{\star \mathrm{op}}$, where $SU_2^{\star \mathrm{op}}$ is a solvable Lie group. We obtain it of course from the Lie bialgebra duality.

As a corollary, by the algebraic group version of the bicrossproduct Hopf algebras at the end of the last lecture, we obtain Hopf algebras

$$\mathbb{C}[U^{\star \mathrm{op}}] \blacktriangleright\!\!\!\triangleleft U(\mathfrak{u})$$

from the Iwasawa decomposition associated to the standard compact real forms of all complex semisimple Lie algebras $\mathfrak{g}$.

Example 22.10 *The bicrossproduct Hopf algebra* $\mathbb{C}(SU_2^{\star \mathrm{op}}) \blacktriangleright\!\!\!\triangleleft U(su_2)$ *is* $k\langle e_i, x_i, g^{-1} \rangle$, *where* $i = 1, 2, 3$ *and* $g = x_3 + 1$, *modulo the relations*

$$[x_i, x_j] = 0, \quad [e_i, e_j] = \epsilon_{ijk} e_k, \quad [e_i, x_j] = \epsilon_{ijk} x_k - \frac{1}{2} \epsilon_{ij3} g^{-1} \sum_i x_i^2,$$

where ϵ_{ijk} is totally antisymmetric and $\epsilon_{123} = 1$. The coalgebra is

$$\Delta e_i = e_i \otimes g^{-1} + e_3 \otimes g^{-1} x_i + 1 \otimes e_i, \quad \Delta x_i = x_i \otimes 1 + g \otimes x_i$$

and $\epsilon(e_i) = \epsilon(x_i) = 0$.

Proof The Lie group $SU_2^{*\mathrm{op}}$ has an algebraic model $\mathbb{C}[SU_2^{*\mathrm{op}}]$ of its algebra of functions, which appears as a sub-Hopf algebra in the sequence

$$0 \to \mathbb{C}[SU_2^{*\mathrm{op}}] \to \mathbb{C}(SU_2^{*\mathrm{op}}) \blacktriangleright\!\!\triangleleft U(su_2) \to U(su_2) \to 0.$$

Its coproduct corresponds to the solvable group law. The action of $U(su_2)$ on it corresponds to the vector fields in the exponentiation of the Lie algebra actions to Lie group actions. The coaction β of $\mathbb{C}[SU_2^{*\mathrm{op}}]$ is also obtained from this exponentiation. One finds:

$$e_i \triangleright x_j = \epsilon_{ijk}\left(x_k - \frac{1}{2}\delta_{k3}g^{-1}\sum_i x_i^2\right), \quad \beta(e_i) = e_i \otimes g^{-1} + e_3 \otimes g^{-1} x_i.$$

$\square$

Thus we have a large class of bicrossproduct quantum groups (at least one for each complex semisimple Lie algebra) quite different from the $U_q(\mathfrak{g})$. Their relationship with the latter remains little explored. [Such bicrossproduct Hopf algebras have been applied to quantum gravity and models of noncommutative spacetime.]

23

Poisson geometry. Noncommutative bundles. q-Sphere

We are now ready to outline the results in Poisson geometry coming out of quantum groups. The general setting is, say, with smooth real manifolds and smooth maps between them, but we will soon specialise to Lie group manifolds. A Poisson manifold is:

1. A manifold M.
2. A Lie algebra structure $\{\ ,\ \}$ on $C^\infty(M)$.
3. $\{f,\ \}$ is a derivation for all $f \in C^\infty(M)$.

The antisymmetry means that the derivation property also holds in the other input. Hence it is natural to obtain $\{\ ,\ \}$ as the action of an antisymmetric 2-tensor field γ on M,

$$\{f,g\} = \langle \gamma, \mathrm{d}f \otimes \mathrm{d}g \rangle = \gamma^{ij} \partial_i(f) \partial_j(g), \quad \forall f, g \in C^\infty(M),$$

where the second expression is in local coordinates with corresponding partial derivatives ∂_i. In terms of this tensor field, the remaining condition for a Poisson bracket is

$$\sum_a \gamma^{ia} \partial_a \gamma^{jk} + \gamma^{ja} \partial_a \gamma^{ki} + \gamma^{ka} \partial_a \gamma^{ij} = 0$$

for the Jacobi identity. Traditionally, mathematicians have concentrated on Poisson brackets where the 2-tensor γ is everywhere an invertible matrix. In this case, the corresponding condition for $\omega = \gamma^{-1}$, viewed as an antisymmetric 2-cotensor field (a 2-form), is $\mathrm{d}\omega = 0$ in the de Rham cohomology. This is a symplectic manifold. One of the lessons of the theory of Lie bialgebras, however, is that one should not focus too soon on this special invertible case – the Poisson brackets of particular interest to us will not be invertible.

We now focus on Poisson structures on Lie group manifolds. In this case it is natural to define vector fields as generated by right translation from the tangent space at e, which we identify with $\mathfrak{g}$ (the Lie algebra of G). Writing $R_u(v) = vu$ for all $u, v \in G$, the vector field $R_*(x)$ associated to $x \in \mathfrak{g}$ has the value $R_{u*}(x)$ at $u \in G$, where R_{u*} is the derivative of R_u. In a similar way, writing

$$\gamma(u) \equiv R_{u*}(D(u)), \quad D : G \to \mathfrak{g} \otimes \mathfrak{g},$$

the axioms of a Poisson structure in terms of the $\mathfrak{g} \otimes \mathfrak{g}$-valued function D become

$$\tau(D(u)) = -D(u),$$

$$\sum D(u)^{(1)} \otimes R_*(D(u)^{(2)})(D)(u)$$
$$- \frac{1}{2} \sum D(u)^{(1)} \otimes \mathrm{ad}_{D(u)^{(2)}}(D(u)) + \text{ cyclic} = 0,$$

where $D(u) = \sum D(u)^{(1)} \otimes D(u)^{(2)}$ and the vector field $R_*(D(u)^{(2)})$ is applied to D as function on G. Also, $+$ cyclic means to add rotations in $\mathfrak{g} \otimes \mathfrak{g} \otimes \mathfrak{g}$. We will use a similar notation L_u for left translation by $u \in G$.

Definition 23.1 *A 'Poisson–Lie group' is a Lie group equipped with a Poisson bracket such that the product map $G \times G \to G$ is a Poisson map, where $G \times G$ has the direct product structure. Explicitly,*

$$\{f, g\}(uv) = \{f \circ L_u, g \circ L_u\}(v) + \{f \circ R_v, g \circ R_v\}(u), \quad \forall u, v \in G.$$

The notion is due to Drinfeld. In terms of D the additional requirement for a Poisson–Lie group is easily found as

$$D(uv) = \mathrm{Ad}_u(D(v)) + D(u), \quad \forall u, v \in G,$$

where Ad is the action of G on $\mathfrak{g} \otimes \mathfrak{g}$ in the usual way (on both factors of the tensor product). This says that $D \in Z^1_{\mathrm{Ad}}(G, \mathfrak{g} \otimes \mathfrak{g})$ in the complex for Lie group cohomology.

Theorem 23.2 *If $(\mathfrak{g}, \delta)$ is a Lie bialgebra, then the associated connected and simply connected Lie group G is a Poisson–Lie group by exponentiating δ. Conversely, if (G, D) is a Poisson–Lie group, then its Lie algebra is a Lie bialgebra by differentiating D at the identity.*

Proof This is general feature of the correspondence between Lie algebra and Lie group cocycles. In our case, the Lie algebra cocycle $\delta \in Z^1_{\mathrm{ad}}(\mathfrak{g}, \mathfrak{g} \otimes \mathfrak{g})$ extends to a Lie group cocycle $D \in Z^1_{\mathrm{Ad}}(G, \mathfrak{g} \otimes \mathfrak{g})$. We then check that D obeys the condition for a Poisson structure. In the reverse direction, we differentiate D at the identity to obtain δ. $\square$

By this correspondence, the entire theory of Lie bialgebras exponentiates to a theory of Poisson–Lie groups. For example, we have the notion of quasitriangular Poisson–Lie group, etc. In this case D has a particularly nice form:

Corollary 23.3 *If* $(\mathfrak{g}, r, [\ ,\ ])$ *is a quasitriangular Lie bialgebra, then*

$$D(u) = \mathrm{Ad}_u(r) - r.$$

Proof In the quasitriangular case, $\delta = \partial r$ in the Lie algebra cohomology. Regarding r instead as a 0-cochain in G (with values in $\mathfrak{g} \otimes \mathfrak{g}$), it follows that $D = \partial r$ in the group cohomology. This is the equation shown. $\square$

There is a lot more of the theory which we could develop in this setting. Also, the above is stated for real $(\mathfrak{g}, \delta)$; for half-real forms we define instead $\gamma = \imath R_* D$. We content ourselves here with an example.

Example 23.4 SU_2 *is a Poisson–Lie group with Poisson bracket induced by* su_2 *with its standard quasitriangular Lie bialgebra structure. In terms of the complex matrix coordinates* a, b, c, d *(with* $ad - bc = 1$*), it is*

$$\{a, b\} = -\frac{1}{2}ab, \quad \{a, c\} = -\frac{1}{2}ac, \quad \{a, d\} = -cb,$$

$$\{b, c\} = 0, \quad \{b, d\} = -\frac{1}{2}bd, \quad \{c, d\} = -\frac{1}{2}cd.$$

Proof This follows from the definitions above. However, there are also 'r-matrix methods' for this kind of computation. [When we work with coordinates $t^i{}_j$ given by the matrix entries, the vector fields obtained by translation of Lie algebra elements have a particularly simple form in terms of the corresponding matrix representation ρ of the Lie algebra. One finds then

$$\{t^i{}_j, t^k{}_l\} = \sum_{a,b} t^i{}_a t^k{}_b r^a{}_j{}^b{}_l - r^i{}_a{}^k{}_b t^a{}_j t^b{}_l,$$

where $r \in M_n \otimes M_n$ is the image of the quasitriangular structure under $\rho \otimes \rho$.] $\square$

One may consider the quantum group $SL_q(2)$ as the quantisation of this Poisson bracket. More precisely, we would need the real form $SU_q(2)$

of $SL_q(2)$ for the real-geometrical picture. We have not discussed real forms of Hopf algebras over $\mathbb{C}$ but the right notion is that of a Hopf *-algebra, as covered in the third Problems set. Aside from this, and regarding $q = e^{\frac{t}{2}}$, we have $ab - ba = t\{a, b\} + \cdots$, etc. to lowest order in t, which is the conventional notion of 'quantisation' of a Poisson bracket.

We start now a completely different topic. Whereas Poisson geometry is the 'semiclassical' geometry of quantum groups, where the lowest order deformation in the quantum group is referred back to classical geometry as additional structure on it, we want to turn now to the fully 'quantum' geometry associated to quantum groups.

We need first the notion of quotient space. In our arrow-reversed language (where we work with algebras thought of as like functions on spaces rather than spaces themselves), it means the invariant subalgebra

$$A^H = \{a \in A \mid \beta(a) = a \otimes 1\}$$

which exists whenever A is a (say) right H-comodule algebra under a coaction β. This is the analogue of functions on the total space that are invariant under a group action. A principal bundle in differential geometry involves such a quotient and also a 'local triviality' condition. Remarkably, the latter can be replaced in practice by something entirely algebraic as well.

Definition 23.5 *A 'quantum principal bundle' (P, H, β) means*

  1. P a right H-comodule algebra under β; let $M = P^H$.

  2. The map ver : $P \otimes_M P \to P \otimes H$ *defined by* ver$(u \otimes_M v) = u\beta(v)$ *for all $u, v \in P$ is a linear isomorphism.*

The second condition here is called the *Hopf–Galois* condition and is important also in the application of Hopf algebras to field extensions and Galois theory. In the present discussion it has a differential geometric meaning, which we explain next.

Definition 23.6 *Let A be an algebra. A first order 'differential calculus' (Ω^1, d) over it means*

  1. Ω^1 a A-bimodule.

  2. A linear map $\mathrm{d} : A \to \Omega^1$ *(the* exterior derivative*) such that*

$$\mathrm{d}(ab) = (\mathrm{d}a)b + a\mathrm{d}b, \quad \forall a, b \in A.$$

  3. $\Omega^1 = \mathrm{span}\{a\mathrm{d}b \mid a, b \in A\}$.

This is more or less the minimum that one could require for an abstract notion of 'differentials' – one should be able to multiply them from the left and from the right by 'functions' (elements of A), which is then enough to have a well-defined Leibniz rule as shown. In usual algebraic geometry one would assume that the left and right modules coincide, but this is not reasonable to impose when our algebras are noncommutative.

Example 23.7 *The universal differential calculus $\Omega^1 A$ is given by*

1. $\Omega^1 A = \ker m \subseteq A \otimes A$ (the kernel of the product map).

2. $\mathrm{d} : A \to \Omega^1 A$ defined by $\mathrm{d}a = 1 \otimes a - a \otimes 1$ for $a \in A$.

Moreover, any other differential calculus is a quotient of this, $\Omega^1 = \Omega^1 A / \mathcal{N}$, for some subbimodule $\mathcal{N} \subseteq \Omega^1 A$. Its exterior derivative is that of $\Omega^1 A$ followed by the projection to the quotient.

Proof It is elementary to check that $\Omega^1 A$ is indeed a differential calculus. That it is universal follows from the surjectivity axiom 3. Universal here means with the obvious notion of morphisms between calculi, namely bimodule maps that form a commutative triangle with the exterior derivatives. □

Before moving on, let us note that $\Omega^1 A$ extends to an entire complex $\Omega A = \bigoplus_n \Omega^n A$ where $\Omega^n A \subset A^{\otimes(n+1)}$ is the joint kernel of all the product maps between adjacent copies of A in the tensor product. One has

$$\mathrm{d} : \Omega^n A \to \Omega^{n+1} A,$$

$$\mathrm{d}(a_0 \otimes \cdots \otimes a_n) = \sum_{i=0}^{n+1} (-1)^i a_0 \otimes \cdots \otimes a_{i-1} \otimes 1 \otimes a_i \otimes \cdots \otimes a_n.$$

There is also a product

$$(a_0 \otimes \cdots \otimes a_n)(b_0 \otimes \cdots \otimes b_m) = (a_0 \otimes \cdots \otimes a_n b_0 \otimes \cdots \otimes b_m).$$

We can now understand the geometric meaning of the Hopf–Galois condition. The map $\mathrm{v\tilde{e}r}(u \otimes v) = u\beta(v)$ for $u, v \in P$ (from which ver was obtained) restricts to a map

$$\mathrm{v\tilde{e}r} : \Omega^1 P \to P \otimes \ker \epsilon.$$

Evaluating against elements of H^* would give maps $\Omega^1 P \to P$ for each such element, which play the role geometrically of the vertical vector fields on P generated (in the classical case) by action of the Lie algebra of the structure group of the bundle. This is the meaning of $\mathrm{v\tilde{e}r}$.

On the other hand, $P(\Omega^1 M)P \subseteq \Omega^1 P$ are the 1-forms on the base viewed on P [classically they would be by pull-back along the bundle projection]. Because of the form of $\Omega^1 M$ we can also write $P(\Omega^1 M)P = P(dM)P$, and thereby interpret the Hopf–Galois condition as the exactness of the sequence

$$0 \to P(\Omega^1 M)P \to \Omega^1 P \xrightarrow{\text{v\v{e}r}} P \otimes \ker \epsilon \to 0.$$

The surjectivity of v\v{e}r corresponds classically to the freeness of the action, while the middle condition is that the forms from the base M are exactly the 'horizontal forms' in the kernel of v\v{e}r. This replaces the concept of the existence of a local trivialisation, which would normally be used to prove such things in classical differential geometry.

Example 23.8 *Over* $\mathbb{C}$, *let* $P = SL_q(2)$ *and* $H = \mathbb{C}\mathbb{Z} = \mathbb{C}[g, g^{-1}]$ *with coproduct* $\Delta g = g \otimes g$. *There is a Hopf algebra surjection*

$$\pi : SL_q(2) \to \mathbb{C}[g, g^{-1}], \quad \pi \begin{pmatrix} a & b \\ c & d \end{pmatrix} \to \begin{pmatrix} g & 0 \\ 0 & g^{-1} \end{pmatrix}$$

which induces a coaction $SL_q(2) \to SL_q(2) \otimes \mathbb{C}\mathbb{Z}$ *as* $\beta = (\mathrm{id} \otimes \pi) \circ \Delta$. *Explicitly, it is*

$$\beta \begin{pmatrix} a & b \\ c & d \end{pmatrix} = \begin{pmatrix} a \otimes g & b \otimes g^{-1} \\ c \otimes g & d \otimes g^{-1} \end{pmatrix}.$$

We define the 'quantum sphere' as the invariant subalgebra

$$S_q^2 = SL_q(2)^{\mathbb{C}\mathbb{Z}}$$

generated by the elements $b_3 = ad$, $b_+ = cd$ *and* $b_- = ab$. *Thus,* S_q^2 *is* $\mathbb{C}\langle b_3, b_+, b_- \rangle$ *modulo the relations*

$$b_{\pm} b_3 = q^{\pm 2} b_3 b_{\pm} + (1 - q^{\pm 2}) b_{\pm},$$

$$q^2 b_- b_+ = q^{-2} b_+ b_- + (q - q^{-1})(b_3 - 1),$$

$$b_3^2 = b_3 + q b_- b_+.$$

Proof It is easy to see that π is a Hopf algebra map, hence β is a coaction and S_q^2 as defined a subalgebra. The elements $b_3, b_{\pm}$ are fixed under β and in fact generate the subalgebra. Hence S_q^2 can be defined as the free algebra with these generators and the inherited relations as shown. Note that the first set of relations of S_q^2 are 'q-commutativity' relations in the sense that they become that $b_3, b_\pm$ commute when $q = 1$. The remaining relation becomes when $q = 1$,

$$x^2 + y^2 + z^2 = \frac{1}{4}$$

in terms of x, y, z defined by $b_\pm = \pm(x \pm iy)$ and $b_3 = z + \frac{1}{2}$. I.e. these are complex coordinates for the q-sphere. One can make the above constructions over a general field k as well.   □

The real-geometrical motivation for this example is of course the Hopf fibration

$$S^2 = SU_2/U(1)$$

where $U(1) \subset SU_2$ by the diagonal embedding and the action of $U(1)$ on SU_2 is the right action of SU_2 on itself pulled back along the inclusion. We do the same thing above with arrows reversed and in an algebraic setting, with $\mathbb{C}\mathbb{Z}$ as an algebraic model of the coordinate algebra of $U(1)$. Classically, SU_2 is a $U(1)$-bundle in this way over the sphere, and has a canonical connection. This is also true in the quantum case as we will see in the next lecture.

24

Connections. q-Monopole. Nonuniversal differentials

In this last lecture we finish up our brief introduction to the 'quantum' geometry of which quantum groups are a part. If the theory of quantum groups is young, this is even younger, but it is one of the main motivations for the interest of quantum groups in physics, for example.

We have already defined quantum principal bundles (P, H, β) in the last lecture, where P is a right H-comodule algebra with coaction β and

$$0 \to P(\Omega^1 M)P \to \Omega^1 P \xrightarrow{\text{v\~er}} P \otimes \ker \epsilon \to 0$$

is exact. Here $M = P^H$ plays the role of (the coordinate ring of) the 'base' of the bundle, P the 'total space' and H the 'structure group'. We define a connection as an equivariant splitting of this sequence, i.e. a choice of equivariant complement to $P(\Omega^1 M)P \subseteq \Omega^1 P$.

Definition 24.1 *A* connection *on a quantum principal bundle* (P, H, β) *is* $\Pi : \Omega^1 P \to \Omega^1 P$ *where*

1. $\Pi^2 = \Pi$ *and* $\ker \Pi = P(\Omega^1 M)P$.

2. Π *is a left P-module map.*

3. Π *commutes with the coaction of H, where H coacts on* $\Omega^1 P \subset P \otimes P$ *by the tensor product coaction.*

As in classical geometry, there is a correspondence between such projections and connection 1-forms:

Proposition 24.2 *There is a 1-1 correspondence between connections* Π *and* $\omega : \ker \epsilon \to \Omega^1 P$ *such that*

1. $\text{v\~er} \circ \omega = 1 \otimes \text{id}$.

2. ω *commutes with the coaction of H, where H coacts on $\ker \epsilon$ by the right adjoint coaction* Ad_R.

Proof (Sketch) Given ω we define

$$\Pi = (m \otimes \mathrm{id}) \circ (\mathrm{id} \otimes \omega) \circ \widetilde{\mathrm{ver}}$$

and easily verify that we have the desired properties. Conversely, given Π, we define

$$\omega(h) = \Pi \circ \mathrm{ver}^{-1}(1 \otimes h), \quad \forall h \in \ker \epsilon,$$

and check that ω is well-defined and has the right properties, and that the constructions are mutually inverse. The different liftings of ver^{-1} to $\Omega^1 P$ differ by the kernel of Π. $\qquad\square$

One can go on and define the curvature form, associated quantum vector bundles (they are typically projective modules), covariant derivatives, etc. – the usual elements of differential geometry. We limit ourselves to an example, for which we return to the quantum homogeneous space S_q^2 constructed in Example 23.8. There is a general lemma for this kind of example:

Lemma 24.3 *Let $\pi : P \to H$ be a Hopf algebra surjection between Hopf algebras and $\beta = (\mathrm{id} \otimes \pi)\Delta$. Then*

1. (P, H, β) is a quantum principal bundle if the product map $m :$ $\ker \epsilon|_M \otimes P \to \ker \pi$ is a surjection.

2. If $i : \ker \epsilon_H \to \ker \epsilon_P$ obeys $\pi \circ i = \mathrm{id}$ and $(\mathrm{id} \otimes \mathrm{id})\mathrm{Ad} \circ i = (i \otimes \mathrm{id}) \circ \mathrm{Ad}$, then

$$\omega(h) = \sum (Si(h)_{(1)})di(h)_{(2)}, \quad \forall h \in \ker \epsilon$$

is a connection (the 'canonical connection').

Proof (Sketch) The first part is a technical condition to ensure the existence of ver^{-1} (the Hopf–Galois condition), which we skip here. For the second part, we verify covariance of ω as

$$\beta \circ \omega(h) = Si(h)_{(2)}di(h)_{(3)} \otimes \pi(Si(h)_{(1)})\pi \circ i(h)_{(4)}$$
$$= Si(h_{(2)})_{(1)}di(h_{(2)})_{(2)} \otimes (Sh_{(1)})h_{(3)} = \omega(h_{(2)}) \otimes (Sh_{(1)})h_{(3)}$$

for all $h \in \ker \epsilon$. In the second equality we used the covariance property of i. Moreover,

$$\widetilde{\mathrm{ver}} \circ \omega(h) = (Si(h)_{(1)})i(h)_{(2)} \otimes \pi(i(h)_{(3)}) - (Si(h)_{(1)})i(h)_{(2)} \otimes 1$$
$$= 1 \otimes \pi(i(h)) = 1 \otimes h$$

for all $h \in \ker \epsilon$. We used the explicit form of d. $\qquad\square$

For classical homogeneous spaces the existence of a canonical connection is based on local triviality; one builds it near the identity and then translates it around to other coordinate patches. In our case, we obtain the corresponding result by entirely algebraic methods, which is a new proof even for $q = 1$:

Example 24.4 *The quantum Hopf fibration $SL_q(2) \to \mathbb{C}[g, g^{-1}]$ with base S_q^2 is a quantum principal bundle, and has canonical connection (the q-monopole) defined by*

$$i(g^n - 1) = a^n, \quad i(g^{-n} - 1) = d^n.$$

Explicitly,

$$\omega(g - 1) = dda - qbdc.$$

Proof (Sketch) we verify the conditions in the lemma. It is then elementary to compute ω. □

When $q = 1$ we recover the canonical $U(1)$ connection (the charge 1 monopole) over the sphere. This is a nontrivial connection with nontrivial Chern class etc.

Finally, all of the above was with the universal differential calculus. Such calculi are used in algebraic topology but are much bigger than their classical counterparts even when the algebra is commutative. We now conclude with an introduction to the more interesting nonuniversal differential calculi. Actually, even for an ordinary manifold there is not a unique differential structure (although one tends to use standard ones). The nonuniqueness is even more pronounced in the quantum case where we have weaker axioms (different left and right actions on the differentials). In the classical case of a Lie group the situation is better and one has a unique translation-invariant differential calculus. We likewise have a better situation in the quantum group case, although usually without uniqueness.

Definition 24.5 *A differential calculus Ω^1 on a Hopf algebra H is bicovariant if*

1. Ω^1 is a bicomodule, by left and right coactions $\beta_L : \Omega^1 \to H \otimes \Omega^1$ and $\beta_R : \Omega^1 \to \Omega^1 \otimes H$.

2. β_L, β_R are bimodule maps, where $H \otimes \Omega^1$ and $\Omega^1 \otimes H$ have the tensor product action and H acts on itself by multiplication.

3. d is a bicomodule map, where H coacts on itself by Δ.

A bicomodule is just like a bimodule with arrows reversed (the coactions commute). Also note that the above definition of a bicovariant calculus is completely left–right symmetric. However, for the following classification theorem we have to make a choice to emphasise left or right comodules. To fit with conventions earlier in the course we chose left (in the theory of principal bundles one might prefer the right comodule setting, for example).

Lemma 24.6 *Let H be a Hopf algebra. Then*

1. $\Omega^1 H \cong \ker \epsilon \otimes H$ by $h \otimes g \mapsto \sum h_{(1)} \otimes h_{(2)} g$ for all $h \otimes g \in \Omega^1 H$.

2. $\ker \epsilon \in {}^H_H \mathcal{M}$, the braided category of crossed modules, by the left regular action and $\mathrm{Ad}_L(h) = \sum h_{(1)} Sh_{(3)} \otimes h_{(2)}$.

Proof This is elementary. For the first part, we have a similar isomorphism $H \otimes H \cong H \otimes H$, which we restrict to $\Omega^1 H = \ker m$. The inverse is $h \otimes g \mapsto h_{(1)} \otimes Sh_{(2)} g$. For the second part, we already know from Corollary 10.5 that $H \in {}^H_H \mathcal{M}$ by multiplication and Ad, and we restrict this to $\ker \epsilon$. $\square$

Theorem 24.7 *Bicovariant Ω^1 are in 1-1 correspondence with quotient objects Λ^1 of $\ker \epsilon \in {}^H_H \mathcal{M}$ (i.e. with Ad_L-stable left ideals contained in $\ker \epsilon$.)*

Proof (Sketch) A quotient object Λ^1 is by definition of the form $\Lambda^1 = \ker \epsilon / \mathcal{M}$ for some $\mathcal{M}$ which is a left ideal (to be stable under the left action) and stable under Ad_L in the sense $\mathrm{Ad}_L(\mathcal{M}) \subseteq H \otimes \mathcal{M}$. Given such an Λ^1, we define

$$\Omega^1 = \Lambda^1 \otimes H, \quad \mathrm{d}h = (\pi \otimes \mathrm{id})(\Delta h - 1 \otimes h), \quad \forall h \in H,$$

where $\pi : \ker \epsilon \to \Lambda^1$ is the canonical projection. The right (co)module structures on Ω^1 are those of H alone by right (co)multiplication. The left (co)module structures are the tensor product of those on Λ^1 as an object of ${}^H_H \mathcal{M}$ (as inherited from $\ker \epsilon$ in the above lemma) and those on H by left (co)multiplication. The map d shown is $\mathrm{d}h = h \otimes 1 - 1 \otimes h \in \Omega^1 H$ mapped under the isomorphism in the lemma and projected to $\Lambda^1 \otimes H$. (This is actually $-\mathrm{d}$ in our previous conventions above.) In the converse direction, we know that $\Omega^1 = \Omega^1 H / \mathcal{N}$ for some $\mathcal{N}$ and show that for a bicovariant calculus, $\mathcal{N}$ has the form $\mathcal{M} \otimes 1$ under the isomorphism in the lemma, for some Ad_L-stable left ideal $\mathcal{M}$. Geometrically, one should understand Λ^1 as the space of right-invariant 1-forms associated to any calculus Ω^1. $\qquad\qquad\qquad\qquad\qquad\qquad\qquad\qquad\qquad\qquad\qquad\square$

Before giving a couple of concrete examples of a complete classification, let us note that there is an obvious notion of morphisms between bicovariant differential calculi – maps commuting with the (co)module structures and forming a commutative triangle with d. One says that a calculus is *irreducible* (or 'coirreducible' would be a more precise term) if it has no proper quotients.

Example 24.8 *For $H = k[x]$ with $\Delta x = x \otimes 1 + 1 \otimes x$, the irreducible bicovariant Ω^1 are in 1-1 correspondence with irreducible monic polynomials $m \in k[x]$, and take the form $\Omega^1 = k_\lambda[x]$ where $k_\lambda = k[\lambda]/\langle m \rangle$ is the corresponding separable field extension. The bimodule structures and d are*

$$(f\phi)(\lambda, x) = f(x + \lambda)\phi(\lambda, x), \quad (\phi f)(\lambda, x) = \phi(\lambda, x)f(x),$$

$$\mathrm{d}f = \tfrac{f(x+\lambda)-f(x)}{\lambda}$$

for all $f \in k[x]$, $\phi \in k_\lambda[x]$.

Proof Here Ad_L is trivial so, by the theorem, bicovariant differential calculi on $k[x]$ are in 1-1 correspondence with ideals $\mathcal{M} \subset \ker \epsilon = \langle x \rangle$ (the ideal generated by x in $k[x]$). Since $k[x]$ is a PID, the ideal $\mathcal{M}$ is generated by a polynomial. Since $\mathcal{M} \subset \ker \epsilon$, this polynomial is divisible by x, i.e. $\mathcal{M} = \langle xm \rangle$ for some m. In this way, irreducible calculi correspond to m irreducible and monic.

We identify the corresponding $\Lambda^1 = \langle x \rangle / \langle xm \rangle$ with $k_\lambda = k[\lambda]$ by

$$\Lambda^1 \cong k_\lambda, \quad xf(x) \mapsto f(\lambda).$$

Under this identification, $\Omega^1 = \Lambda^1 \otimes k[x] \cong k_\lambda[x]$. The action from the right is by the inclusion $k[x] \subset k_\lambda[x]$. The action from the left is by

$$f(x) \cdot x^m \otimes x^n = f(x \otimes 1 + 1 \otimes x)x^m \otimes x^n.$$

as the tensor product action. Hence $f(x) \cdot \lambda^{m-1} x^n = f(\lambda + x)\lambda^{m-1} x^n$ under our identification. These expressions are modulo $\langle xm(x) \rangle$ or $\langle m(\lambda) \rangle$.

Finally, we compute $\mathrm{d}f = f(x \otimes 1 + 1 \otimes x) - 1 \otimes f(x)$ modulo $\langle xm \rangle$ in the first tensor factor. Under our isomorphism this is the expression stated, modulo $\langle m(\lambda) \rangle$. Note that $dx = x \otimes 1$ modulo $\langle xm \rangle$ corresponds to $1 \in k_\lambda[x]$ under our identification. $\qquad\qquad\qquad\qquad\qquad\qquad\square$

For a concrete example, the bicovariant calculi on $\mathbb{C}[x]$ are parameterised by $\lambda_0 \in \mathbb{C}$ (say). Here $m(\lambda) = \lambda - \lambda_0$ and $\pi(\lambda) = \lambda_0$. Hence,

$$df = dx \, \frac{f(x + \lambda_0) - f(x)}{\lambda_0}$$

(understood as dx times the coefficient of λ in $f(x+\lambda)$ followed by setting $\lambda = \lambda_0$). This is a complete classification and only the case $\lambda_0 = 0$ is the standard translation-invariant calculus discovered by Isaac Newton; for $\lambda_0 \neq 0$ one has $f dg \neq (dg)f$ for polynomials f, g (a bimodule structure with different left and right actions), but it is a perfectly good quantum differential calculus and actually much better behaved.

Example 24.9 *For $H = k(G)$ on a finite group G, the irreducible Ω^1 are in 1-1 correspondence with nontrivial conjugacy classes $C \subset G$. The 1-forms $e_a = \sum_{g \in G}(d\delta_{ag})\delta_g$ for $a \in C$ form a basis of Λ^1 and*

$$f.e_a = e_a L_a^*(f), \quad df = \sum_{a \in C} e_a(L_a^*(f) - f).$$

Proof This is immediate from the theorem above. For Λ^1 we set to zero all delta-functions except $\{\delta_a\}_{a \in C}$. These project to our basis $\{e_a\}$, which we identify in terms of d. Then Ω^1 is a free right $k(G)$-module with basis $\{e_a\}$ and we give the left module and d in these terms. Here $L_a^*(f) = f(a(\ ))$. $\qquad\qquad\qquad\qquad\qquad\qquad\qquad\qquad\qquad\square$

To round off the course we note that the dual of Λ^1 is typically a braided-Lie algebra as in Figure 14.5. In the above example it has basis $\{x_a\}_{a \in C}$ with

$$\Psi(x_a \otimes x_b) = x_b \otimes x_a, \quad [x_a, x_b] = x_{aba^{-1}}, \quad \Delta x_a = x_a \otimes x_a, \quad \epsilon x_a = 1.$$

Meanwhile for $SL_q(2)$ the Ω^1 essentially correspond to irreducible representations of $U_q(sl_2)$. The 2-dimensional one leads to a 4-dimensional Λ^1. Its dual is the braided-Lie algebra $\widetilde{sl}_{2,q}$.

Problems

* questions are optional.

1. If a Hopf algebra is commutative, show that $S^2 = \mathrm{id}$.

2.* In any Hopf algebra, show that $\Delta \circ S = \tau \circ (S \otimes S) \circ \Delta$ and $\epsilon \circ S = \epsilon$, where τ is the twist map (i.e. an anticoalgebra map).

3. In any Hopf algebra, show that

$$\mathrm{Ad}_R(h) = \sum h_{(2)} \otimes (Sh_{(1)})h_{(3)}$$

makes H into a right H-comodule coalgebra. You may assume the result of Q2.

4. If V, W are right comodules, show that their tensor product is also a comodule in a natural way.

5. Show that if H acts from the left on a vector space V then it also acts on V^*, by $(h{\triangleright}f)(v) = f((Sh){\triangleright}v)$ for all $h \in H$, $v \in V$ and $f \in V^*$.

6. Show that $\alpha \in H$ is central iff $\mathrm{Ad}_h(\alpha) = \epsilon(h)\alpha$ for all $h \in H$.

7. Compute the left action $\mathrm{Ad}_h(g) = \sum h_{(1)}gSh_{(2)}$ for (i) kG (G a finite group) (ii) $U(\mathfrak{g})$ (give the action of the Lie algebra $\mathfrak{g}$) (iii) $U_q(b_+)$.

8.* Let $[\ ,\ ] : H \otimes H \to H$ be the bilinear map $[h, g] \equiv \mathrm{Ad}_h(g)$ (a convenient notation). Show that

$$[x, [y, z]] = \sum [[x_{(1)}, y], [x_{(2)}, z]], \quad \forall x, y, z \in H$$

(the 'pentagonal Jacobi identity'). What does it reduce to on the generators when $H = U(\mathfrak{g})$?

9. Let G be a finite group. Show that an action of $k(G)$ on a vector space is the same thing as a G-grading. What does a module algebra under $k(G)$ mean? (Hint: consider the action of the Kronecker functions δ_g for $g \in G$.)

159

10. Compute the left coregular action $R^*_\phi(h) = \sum h_{(1)} \langle \phi, h_{(2)} \rangle$ for (i) $k(G)$ acting on kG (ii) kG acting on $k(G)$ (iii) $U(\mathfrak{g})$ acting on $\mathbb{C}[G]$ (iv) $U_q(b_+)$ paired with itself (in (iii)–(iv) the action of generators is enough). (Hint for (iv): use the module algebra property.)

11. Let $q \in k^*$. On polynomials $k[x]$, show that the operation $\partial_q(x^n) = [n]_q x^{n-1}$ is a q-derivation in the sense

$$\partial_q(fg) = (\partial_q f)g + f(qx)\partial_q g, \quad \forall f, g \in k[x].$$

Hence or otherwise, show that $k[x]$ is a $U_{q^{-1}}(b_+)$-module algebra where X acts by ∂_q.

Check that the same holds on formal powerseries $k[\![x]\!]$. Assuming that $[n]_q \neq 0$ for all $n \in \mathbb{N}$ (one says that q is 'generic'), observe that $\partial_q e_q^{\lambda x} = \lambda e_q^{\lambda x}$ for all $\lambda \in k$. Hence or otherwise, show that e_q^x has inverse $e_{q^{-1}}^{-x}$.

12.* Obtain a description of the self-duality pairing $\langle X^m g^n, X^r g^s \rangle$ of $U_q(b_+)$ with itself in terms of ∂_q acting on $k[X]$.

13. Show that $M_q(2)$ defined as $k\langle a, b, c, d \rangle$ modulo the relations

$$ca = qac, \quad ba = qab, \quad db = qbd, \quad dc = qcd,$$

$$bc = cb, \quad da - ad = (q - q^{-1})bc$$

is a bialgebra with the matrix coalgebra on the generators $\begin{pmatrix} a & b \\ c & d \end{pmatrix}$. Show that the element $ad - q^{-1}bc$ is central and grouplike in $M_q(2)$.

14.* Let H be a finite-dimensional Hopf algebra and view $\mathcal{R} \in H \otimes H$ as a map $H^* \to H$ sending $\phi \mapsto (\phi \otimes \mathrm{id})(\mathcal{R})$. Cast the quasitriangularity axioms $(\Delta \otimes \mathrm{id})\mathcal{R} = \mathcal{R}_{13}\mathcal{R}_{23}$ and $(\mathrm{id} \otimes \Delta)\mathcal{R} = \mathcal{R}_{13}\mathcal{R}_{12}$ in terms of this map.

15. Show that $\mathbb{C}_{q^2}\mathbb{Z}/n$ is factorisable for q a primitive n'th root of unity and $n > 1$ odd. (Hint: when m runs through $\mathbb{Z}/n$, so does $2m$ because n is odd.)

16.* If H is a bialgebra and $\chi \in H \otimes H$ is invertible and obeys $\chi_{12}(\Delta \otimes \mathrm{id})\chi = \chi_{23}(\mathrm{id} \otimes \Delta)\chi$ and $(\epsilon \otimes \mathrm{id})\chi = 1$, show that H_χ, defined as the same algebra as H but with the new coproduct $\Delta_\chi(h) = \chi(\Delta h)\chi^{-1}$ for all $h \in H$, makes H_χ a bialgebra. This is called the 'twisting operation' among Hopf algebras. What is the equation for χ when $H = k(G)$? (Answer: a group 2-cocycle.)

17. Let $q \in k^*$. Define the q-binomial coefficients inductively by

$\left[\begin{smallmatrix} n \\ m \end{smallmatrix}\right]_q = q^{n-m} \left[\begin{smallmatrix} n-1 \\ m-1 \end{smallmatrix}\right]_q + \left[\begin{smallmatrix} n-1 \\ m \end{smallmatrix}\right]_q$ and $\left[\begin{smallmatrix} n \\ 0 \end{smallmatrix}\right]_q = 1$ (and $\left[\begin{smallmatrix} n \\ m \end{smallmatrix}\right]_q = 0$ when $m > n$). Show that $[m]_q[n-m]_q \left[\begin{smallmatrix} n \\ m \end{smallmatrix}\right]_q = [n]_q$. (Hint: show first that $[m]_q \left[\begin{smallmatrix} n \\ m \end{smallmatrix}\right]_q = \left[\begin{smallmatrix} n-1 \\ m-1 \end{smallmatrix}\right]_q [n]_q$.)

18. Show that $e_q^{A+B} = e_q^A e_q^B$ where A, B obey $BA = qAB$ and are jointly nilpotent in the sense that there exists N such that $A^m B^n = 0$ for all m, n such that $m + n = N$. Assume that $[m]_q \neq 0$ for all $m < N$.

19. In $U_q(sl_2)$, show that the element

$$qg^{-1} + q^{-1}g + (q - q^{-1})^2 EF$$

is central (this is called the q-Casimir).

20.* Let q be a primitive odd n'th root of unity. Show that $\{g^a E^b F^c \mid a, b, c = 0, \ldots, n-1\}$ is a basis of $u_q(sl_2)$.

21.* For H' dually paired with H, the quantum double $H'^{\mathrm{op}} \bowtie H$ is built on the vector space $H' \otimes H$ with the product

$$(\phi \otimes h)(\psi \otimes g) = \sum \psi_{(2)}\phi \otimes h_{(2)}g\langle Sh_{(1)}, \psi_{(1)} \rangle \langle h_{(3)}, \psi_{(3)} \rangle$$

for all $\phi, \psi \in H'$, and $h, g \in H$. Show that this is associative.

Problems II

* questions are optional.

1. Show that $R \in M_n \otimes M_n$ obeys $R_{12}R_{13}R_{23} = R_{23}R_{13}R_{12}$ (the Yang–Baxter equations) iff $\Psi = \tau \circ R \in M_n \otimes M_n$ obeys the braid relations $\Psi_{12}\Psi_{23}\Psi_{12} = \Psi_{23}\Psi_{12}\Psi_{23}$. Here the suffixes refer to the position in $M_n \otimes M_n \otimes M_n$ and τ is the permutation operator $k^n \otimes k^n \to k^n \otimes k^n$ viewed in $M_n \otimes M_n$. (Hint: $\tau_{12}R_{23}\tau_{12} = R_{13}$ etc.)

2. If $H, \mathcal{R}$ is quasitriangular, check that Ψ as in Proposition 10.1 obeys $\Psi_{V \otimes W, Z} = \Psi_{V,Z} \circ \Psi_{W,Z}$ (completing the proof that $_H\mathcal{M}$ is braided).

3.* Let $H, \mathcal{R}$ be a dual quasitriangular Hopf algebra. Show that its category of right comodules is braided via

$$\Psi(v \otimes w) = w^{(\bar{1})} \otimes v^{(\bar{1})}\mathcal{R}(v^{(\bar{2})} \otimes w^{(\bar{2})}).$$

4.* Let B be a braided group in a braided category. Show that if V, W are B-modules in the category, then so is $V \otimes W$ as claimed in Lecture 16. Define a braided B-module algebra as an algebra A in the category and a B-module such that the product (and unit) maps are morphisms. Show that the adjoint action Ad makes B a B-module algebra. (Hint: use diagrams.)

5. What is the braided adjoint action of the braided line $\mathbb{A}_q^1 = k[x]$ on itself, in the category of $k_q\mathbb{Z}$-comodules? Here the coaction is $x \mapsto x \otimes g$ where $k_q\mathbb{Z} = k[g, g^{-1}]$.

6. Show that the braided coproduct of the quantum plane $\mathbb{A}_{q^2}^2$ is necessarily

$$\Delta(x^m y^n) = \sum_{r=0}^{m}\sum_{s=0}^{n} \begin{bmatrix} m \\ r \end{bmatrix}_{q^2} \begin{bmatrix} n \\ s \end{bmatrix}_{q^2} x^r y^s \otimes x^{m-r}y^{n-s}q^{(m-r)s}$$

on general basis elements. (Hint: obtain Δx^m and Δy^n first and then consider the braiding $\Psi(x^{m-r} \otimes y^s)$ implied by $\Psi(x \otimes y) = qy \otimes x$ and functoriality under the product.)

Hence or otherwise, obtain the partial derivatives $\partial_{q,x}$ and $\partial_{q,y}$ stated in Lecture 15, and check that $\partial_{q,y}\partial_{q,x} = q^{-1}\partial_{q,x}\partial_{q,y}$.

7.* Check that if B is a braided group with invertible antipode then

B^{cop} with coproduct $\Psi^{-1} \circ \Delta$ is a braided group in the category with opposite braiding. (Hint: try $S^{\mathrm{cop}} = S^{-1}$.)

8. If A is an H-module algebra, show that $A \rtimes H$ with product $(a \otimes h)(b \otimes g) = a(h_{(1)} \triangleright b) \otimes h_{(2)}g$ for $a, b \in A$ and $h, g \in H$ is associative. If C is a right H-comodule coalgebra, show that $H \blacktriangleright\!\!< C$ is a coalgebra in a natural way.

9.* If H is dual quasitriangular, check that there is a monoidal functor $\mathcal{M}^H \to \mathcal{M}_H^H$ given by

$$(V, \beta) \mapsto (V, \beta, \triangleleft), \quad v \triangleleft h = \sum v^{(1)} \mathcal{R}(v^{(2)}, h), \quad \forall v \in V, \ h \in H.$$

10. Let q be a primitive n'th root of 1. Compute the bosonisation of $k[x]/\langle x^n \rangle$ in the category of $k_q \mathbb{Z}/n$-modules, where $g \triangleright x = qx$.

11. Let $k[y]/\langle y^n \rangle$ be the braided group in the same category as in Q10, where $g \triangleright y = q^{-1}y$. Show that $\mathrm{ev}(y \otimes x) = 1$ extends as pairing of braided groups by

$$\mathrm{ev}(y^a \otimes x^b) = \delta_{a,b}[a]_q!.$$

What is the corresponding coevaluation coev (so that $k[x]/\langle x^n \rangle$ is rigid)?

Problems III

* questions are optional.

1. For $f \in \mathbb{A}_q^1 = k[x]$ the braided line (in the category $k_q\mathbb{Z}$-comodules), prove the braided Taylor's theorem

$$e_q^{x\partial_q} f(y) = f(x+y)$$

in the braided tensor product algebra $k[x]\underline{\otimes}k[y]$. Here ∂_q act in the second factor (the y variable) and not on x. (Hint: the right hand side is $\overline{\Delta f}$.)

2. A Hopf *-algebra is a Hopf algebra over $\mathbb{C}$ which is a *-algebra (i.e. equipped with an antilinear antialgebra map obeying $*^2 = \mathrm{id}$) such that * commutes with Δ, $\epsilon \circ * = {}^{-} \circ \epsilon$ (where $^{-}$ is complex conjugation) and $(S \circ *)^2 = \mathrm{id}$. In the finite-dimensional case, show that H^* is a Hopf *-algebra with $\langle \phi^*, h \rangle = \overline{\langle \phi, (Sh)^* \rangle}$ for all $h \in H, \phi \in H^*$.

3. Check that the following are Hopf *-algebra structures on $U_q(sl_2)$ over $\mathbb{C}$.
   (i) q real, $g^* = g$, $E^* = gF$, $F^* = Eg^{-1}$ (this is called $U_q(su_2)$).
   (ii) q real, $g^* = g$, $E^* = -gF$, $F^* = -Eg^{-1}$ (this is called $U_q(su_{1,1})$).
   (iii) q of modulus 1, $g^* = g$, $E^* = -qE$, $F^* = -q^{-1}F$ (this is called $U_q(sl_2(\mathbb{R}))$).

4.* Let H be a Hopf algebra. Check that $D(H)\cong H^{*\mathrm{op}}\bowtie H$ by mutual coadjoint actions

$$h\triangleleft\phi = \sum h_{(2)}\langle\phi, (Sh_{(1)})h_{(3)}\rangle, \quad h\triangleright\phi = \sum \phi_{(2)}\langle h, (S\phi_{(1)})\phi_{(3)}\rangle$$

for all $h \in H$ and $\phi \in H^*$. Here S denotes the antipode of H or H^*.

5. Let $(\mathfrak{g}, \delta)$ be a finite-dimensional Lie bialgebra. Check that $\mathfrak{g}^*$ is also, with the dual structure maps. Find the dual of the Lie bialgebra sl_2 defined by

$$[H, X_\pm] = \pm 2X_\pm, \quad [X_+, X_-] = H,$$

$$\delta(H) = 0, \quad \delta(X_\pm) = \tfrac{1}{2}(X_\pm \otimes H - H \otimes X_\pm).$$

6.* Check that the fixed subalgebra of $SL_q(2)$ under the coaction $k[g, g^{-1}]$ is the algebra of the q-sphere given in the Lecture 23.

7.* The quantum differential calculus associated to the field extension $\mathbb{R} \subset \mathbb{C}$ is $\Omega^0 = \mathbb{R}[x]$ and $\Omega^1 = \mathbb{C}[x]$. Considering the 1-forms dx and $\omega = x dx - (dx)x$ as a basis of $\mathbb{C}[x]$ as a right $\mathbb{R}[x]$-module, show that the left module structure is

$$x dx = (dx)x + \omega, \quad x\omega = \omega x - dx.$$

Obtain an explicit description for the exterior derivative $d : \mathbb{R}[x] \to \mathbb{C}[x]$.

Bibliography

T. Brzezinski and S. Majid, Quantum group gauge theory on quantum spaces, *Commun. Math. Phys.* **157** (1993) 591–638.

V.G. Drinfeld, Quantum Groups, in *Proceedings of the ICM*, A.M.S. 1987.

G. Lusztig, *Introduction to Quantum Groups*, Birkhäser 1993.

S. Mac Lane, *Categories for the Working Mathematician*, Springer–Verlag 1974.

S. Majid, Algebras and Hopf Algebras in Braided Categories, *Lec. Notes Pure Appl. Math.* **158** (1994) 55–105. Marcel Dekker.

S. Majid, Quantum and braided Lie algebras, *J. Geom. Phys.* **13** (1994) 307–356.

S. Majid, *Foundations of Quantum Group Theory*, C.U.P. 1995

S. Majid, Quantum and braided diffeomorphism groups, *J. Geom. Phys.* **28** (1998) 94–128.

S. Majid, Classification of bicovariant differential calculi, *J. Geom. Phys.* **25** (1998) 119–140.

S. Majid, Double bosonisation of braided groups and the construction of $U_q(\mathfrak{g})$, *Math. Proc. Camb. Phil. Soc.* **125** (1999) 151–192.

M.E. Sweedler, *Hopf Algebras*, Benjamin 1969.

S.L. Woronowicz, Differential calculus on compact matrix pseudogroups (quantum groups), *Commun. Math. Phys.* **122** (1989) 125–170.

Index